CHEMISTRY IN NON-AQUEOUS SOLVENTS

BERNARD TRÉMILLON

CHEMISTRY IN NON-AQUEOUS SOLVENTS

D. REIDEL PUBLISHING COMPANY

DORDRECHT-HOLLAND / BOSTON-U.S.A.

LA CHIMIE EN SOLVANTS NON-AQUEUX
First published by Presses Universitaires de France, Paris, 1971
Translated from the French by N. Corcoran

Library of Congress Catalog Card Number 73–86094

ISBN 90 277 0389 2

Published by D. Reidel Publishing Company
P.O. Box 17, Dordrecht, Holland

Sold and distributed in the U.S.A., Canada, and Mexico
by D. Reidel Publishing Company, Inc.
306 Dartmouth Street, Boston,
Mass. 02116, U.S.A.

Printed in The Netherlands by D. Reidel, Dordrecht

CONTENTS

EDITORIAL NOTE

This book has been translated from *La chimie en solvants non-aqueux* with some small additions and alterations. To facilitate comparison with the original French book, the same system of numbering of sections and equations has been used, and the table entries are given in the same order. In the English translation the tabular matter may sometimes appear in an illogical order, as the alphabetical order of the French original has now disappeared.

INTRODUCTION

Arising no doubt from its pre-eminence as a natural liquid, water has always been considered by chemists as the original solvent in which very varied chemical reactions can take place, both for preparational and for analytical purposes. This explains the very long-standing interest shown in the study of aqueous solutions. In this connection, it must be stressed that the theory of Arrhenius and Ostwald (1887–1894) on electrolytic dissociation, was originally devised solely for solutions in water and that the first true concept of acidity resulting from this is linked to the use of this solvent. The more recent development of numerous physico-chemical measurement methods has made possible an increase of knowledge in this area up to an extremely advanced degree of systematization. Thus today we have available both a very large amount of experimental data, together with very refined methods of deduction and of quantitative treatment of chemical reactions in solution which enable us to make the fullest use of this data.

Nevertheless, it appears quite evident at present that there are numerous chemical processes which cannot take place in water, and that its use as a solvent imposes limitations. In order to overcome these limitations, it was natural that interest should be attracted to solvents other than water and that the new possibilities thus opened up should be explored. We must indeed expect the course of chemical reactions to be profoundly influenced by the solvent forming the reaction medium: by changing it, different products may be obtained starting with the same given reagents, reactions not taking place in water may take place, or vice versa.

The interest presented by these developments is not only at speculative level. There are numerous applications resulting from them, some of which have been realized in industry and are of great importance from the economic point of view. We will quote some significant examples, with the object of showing the wide range concerned.

Organic chemists were no doubt amongst the first to realize the interest of non-aqueous solvents for their work. In particular, one of the most important methods of synthesis in organic chemistry consists of use of organometallic compounds, especially organo-magnesium compounds (Grignard reagents); these are incompatible with water, which decomposes them by hydrolysis. On the other hand, they are soluble in certain organic solvents: diethyl ether, tetrahydrofuran, hexamethylphosphorotriamide, etc. When the solutions so obtained are used to effect transformations, each of these solvents plays a special part effecting the course of the reactions observed.

Biochemists have also been able to make use of certain

solvents. Thus it has been found that numerous proteins (insulin, albumen, haemoglobin, collagen), insoluble in water, are very easily dissolved by anhydrous hydrofluoric acid; the same applies to enzymes, chlorophyll, vitamin B_{12}, which can be subsequently recovered without having lost their biological activity.

Metallurgists make more and more frequent use of non-aqueous solvents for electrolytic treatment of their materials. This is particularly true for aluminium, the most abundant of the metallic elements, which cannot be obtained in the metallic state by an 'aqueous' route, since firstly its most common compound in the natural state, alumina, is insoluble in water, and secondly the metal is too electropositive to be obtained by reduction of the cation Al^{3+} in aqueous solution. The whole aluminium industry is in fact based on the process developed by Hall and Heroult (1886), of electrolytic reduction of alumina solutions in cryolite (Na_3AlF_6) in the molten state, which above 1000 °C acts as a true solvent for Al_2O_3 and enables the molten metal to be separated. The preparation of alkali and alkaline earth metals, electrolytic refining of uranium, etc, are carried out in a similar manner in non-aqueous media (mainly molten salts).

Electrochemists have been led to exploit non-aqueous solvents for still more purposes. The realization of electrochemical generators of high energy yield, especially where the aim is to supply electrical energy to equipment which must be kept light in weight, has led to the development of systems based on light elements, the properties of which prevent the use of aqueous solutions. Hence anodes of lithium or magnesium need, either to operate satisfactorily, or simply to preserve themselves, to be in contact with special solutions, completely free from water, and compatible with the compounds being used. New techniques have also appeared in this domain, such as 'thermal' cells, based on solvents which are liquid at high temperatures, and which are set into operation by a sudden heating causing the electrolyte to fuse.

New techniques in physics are offered by certain inorganic solvents with heavy atoms (such as selenium or phosphorus oxychlorides) for realization and study of liquid lasers. In this connection, search for highly fluorescent lanthanide ion solutions leads to the exclusion of solvents containing light atoms, such as carbon and hydrogen – hence of water and of the usual organic solvents – which favour mechanisms of rapid non-radiative relaxation. It has been found possible to obtain satisfactory results with the above-mentioned solvents, after solving certain problems concerned with solution chemistry.

Lastly, it is in the wide sphere of analytical chemistry that the most extensive and most systematic use of non-aqueous solvents has taken place up to the present. The effect at first investigated was the increase in solubility of the substances to be analyzed. But more fundamental information about properties in solution has permitted the realization of many new possibilities offered by this method for determination of substances which cannot be estimated by aqueous methods (very weak acids or bases in particular).

So, very many chemists are taking an interest today in non-aqueous solutions, since they understand the range of possibilities offered by these latter for resolving their problems. There are no doubt many as yet unsuspected possibilities. Despite the very wide range of knowledge at present available, as is witnessed by abundant literature, it is probable that the practical application of non-aqueous media is only at its beginning.

Rational and effective application demands, much more than description of what takes place in each of the media, establishment of laws of interpretation and general rules allowing predictions to be made. In this more fundamental aspect, the chemistry of non-aqueous solvents started its development at about the beginning of the century; E. C. Franklin in the United States, working on properties in liquid ammonia, P. Walden and A. Hantzsch in Germany with their similar studies of properties in sulphuric acid or in acetonitrile, were amongst the main precursors. But it is only within the last twenty years that fundamental research in this area has been very intense, bringing chemistry in non-aqueous solvents step by step to as advanced a state of systematization as chemistry in aqueous solution.

Our aim here is to explain the general ideas enabling us to understand the problems set, at theoretical levels, by use of non-aqueous solvents for carrying out chemical and electrochemical reactions. It is in no way an exhaustive account, and the particular properties mentioned are only illustrations intended to support the arguments and to make them more explicit. For this same reason, no exhaustive bibliography will be given; only the authors of the work referred to will be quoted.*

The conceptual scheme in accordance with which we have developed our account is as follows. After setting out the different types of solvent which can be used, and the problems of solvating the chemical substances in solution, it will be shown how the ionic theories used for treatment of the properties in aqueous solution are extended to other solvents, with such adaptations as are necessary. Chapters II, III, and IV will deal with this aspect, for acid-base systems, then for other systems of ionic exchange where the solvent can take part, and lastly for oxidizing-reducing systems.

In the last chapter, we have attempted to solve the problem, which is at present without doubt the most

* I take advantage of the opportunity to express my thanks to my collaborators, whose work, sometimes very recent, has provided me with many explanatory examples.

So that the interested reader may obtain further information on each of these topics, a list of the complete references of the papers in which they have been published will be found, in alphabetical order of authors quoted, at the end of this work.

difficult and most fluid, of the correlation between the properties in the different solvents. This is a problem hitherto rarely dealt with in specialized works, but nevertheless essential for an understanding of the role of the solvent and for a considered utilization of non-aqueous media. Resolution of this problem necessitates recourse to considerations on the thermodynamics of solutions, the experimental aspect of which raises difficulties which it has seemed to us to be worth stressing.

Treatment of the different questions has been kept brief by choice, so as to remain within the framework laid down. Certain questions have even been only mentioned and have not received the attention which they perhaps deserved. The selections which have been made have been in accordance with the general conception of the SUP series of books.

We are very grateful to Professor J. Bénard for entrusting us with the task of compiling this work.

CHAPTER I

SOLVENTS AND SOLUTES

I.1. The Different Types of Solvent

Every liquid is a priori capable of use as a solvent, between its solidification temperature and – unless it decomposes beforehand – its boiling point, the temperature of which may by further raised by an increase in pressure. We therefore have a very large number of possible solvents at our disposal, which can be used over a very wide range of temperatures.

We find in practice, however, that there are 'good' and 'bad' solvents, which restricts this generality to some slight extent. Nevertheless, up to now more than three hundred pure liquids have been the subject of investigations as to their properties as solvents. In addition, possibilities are not limited to pure substances only, since we may still consider mixtures of solvents, enabling us not only to extend the range of temperatures used (due to eutectics at the solidification end and to negative azeotropes at the boiling end), but also to change the physico-chemical properties in a continuous manner so realizing a quasi-infinite number of reactive media.

Practically the whole temperature range of chemistry is covered by the solvents used up to the present, both low-temperature chemistry (from below —100 °C up to room temperatures) and high-temperature chemistry (from room temperatures to well above 1000 °C).

On a basis of considerations on structure in the liquid state, it is possible to distinguish several classes of solvent.

1. Molecular solvents

Being constituted from molecules frequently more or less strongly interassociated, molecular solvents are liquid dielectrics, almost non-conductors of electricity (the small conductivity of certain of these, even in the completely pure state, is due to a very small auto-ionization of the molecules; cf. Chapters II and III).

Most molecular solvents are organic compounds: alcohols, ketones, carboxylic acids, amines, amides, nitriles, esters, ethers, hydrocarbons and halogen derivatives, etc. Amongst inorganic molecular solvents other than water we may quote: sulphuric, hydrofluoric and nitric acids, liquid ammonia and sulphur dioxide, phosphorus oxychloride, etc.

These are the solvents most commonly used at or about ordinary temperature. The solvents which can be used at low temperatures (ammonia, sulphur dioxide) are equally suitable. Lastly there are some molecular compounds which can be used as solvents at temperatures higher than normal, on account of their relatively high melting point (m.p.) and sufficient thermal stability (e.g. acetamide, m.p. 81 °C, dimethylsulphone, m.p. 109 °C).

Attempts have frequently been made to connect the influence of the solvent on the chemical properties of the solutes with characteristic physical properties, in particular the dipole moment of the molecules and above all with the dielectric constant (relative permittivity, ε). Table I gives a list of the main molecular solvents, with their temperature range of application and the value of this important constant, the role of which will be seen later.

Apart from this, we have numerous proofs of the influence of the molecular liquid structure on its properties as a solvent. A knowledge of the former also plays an increasing part in the understanding of these latter. It is known that the structure of a liquid is derived from that of a crystal, which is a perfectly ordered rigid edifice; liquefaction corresponds to the appearance of certain disorder features leading to fluidity, cohesion still remaining assured by various interactions between neighbouring molecules: hydrogen bonds in the case of solvents known as 'protonic' solvents (water, alcohols, acids, ammonia and amines, amides, etc.) or association by the intermediary of other 'bridging' atoms (oxygen, halogens) in the case of various 'non-protonic' solvents (halides, oxyhalides, covalent oxides), dipole-dipole interactions, interactions known as 'van der Waals type'.

TABLE I

Principal molecular solvents with their melting points (m.p.), boiling points (b.p.) under atmospheric pressure, and their dielectric constants ε (at ordinary temperature, unless the contrary is stated)

Solvent	m.p.(°C)	b.p.(°C)	ε
Acetamide	82.3	221.2	59.2 (at 100 °C)
Acetone	— 95.3	56.2	20.7
Acetonitrile	— 45.7	80.0	36.2
Acetic acid	16.6	118.5	6.2
Hydrochloric acid	—114.2	— 85.1	11.3 (at—85 °C) 6.3 (at—15 °C)
Hydrofluoric acid	— 83.3	19.5	83.6 at 0 °C)
Formic acid	8.4	100.7	56.1
Nitric acid	— 41.6	82.6	~ 50
Sulphuric acid	— 10.3	270	110
Trifluoroacetic acid	— 15.2	72.4	8.3
Ammonia	— 77.7	— 33.3	22.4 (at—33 °C) 16.9 (at 25 °C)
Sulphur dioxide	— 75.5	— 10.1	15.4 (at 0 °C)
Benzene	5.5	80.1	2.3
Propylene carbonate	— 49.2	241.7	64.2
Chlorobenzene	— 45	132	5.6
Chloroform	— 63.5	61.2	4.8
N, N-Dimethylformamide (DMF)	— 61	153	36.7

Dimethylsulphoxide (DMSO)	18.4	189	48.9
1,4-Dioxane	11.8	101	2.2
Ethanol	—117.3	78.5	24.3
Ethanolamine (aminoethanol)	10.3	170	37.7
Diethyl ether	—116.4	34.6	4.4
Ethylenediamine	8.5	116.5	14.2 (at 30 °C)
Ethylene glycol (1.2-ethanediol)	— 13.2	198	37.7
Formamide	2.5	193	110
Hexamethylphosphorotriamide (HMPA)	7.2	230	30
Isopropanol (2-propanol)	— 89.5	82.4	18.3
Methanol	— 97.8	65.0	32.6
2-Methoxy ethanol (methylcellosolve)	— 85.1	125	16.0
N-Methylacetamide (NMA)	30.5	206	165 (at 40 °C)
Methylisobutylketone (MIBK)	— 80.2	115.9	13.1
Nitrobenzene	5.7	210.8	34.8
Nitromethane	— 28.5	100.8	35.9
Phosphorus oxychloride	1.3	108	13.9
Selenium oxychloride	8.5	178	49.2
Tributyl phosphate (TBP)	— 80.0	289.0	6.8
Pyridine	— 42	115.5	12.3
Carbon tetrachloride	— 23	76.7	2.2
Tetrahydrofuran (THF)	—108.5	66	7.6
Tetramethylenesulphone (sulpholane)	28.8	283	42 (at 30 °C)

Intermolecular associations created by bridging, especially by hydrogen bonds, form true polymers of the unit molecules of the solvent.

The structure of water has especial value, although it is certainly one of the most complex, but naturally the structure most investigated. This liquid appears to be made up of clusters of H_2O molecules associated by hydrogen bonds, forming quasi-crystalline structures immersed in a 'cloud' of non-bonded molecules – in a way rather similar to icebergs. According to spectroscopic measurements (IR, NMR, Raman), these clusters include at ordinary temperatures about 70% of the water molecules, each cluster being a group of about fifty molecules; an increase in temperature naturally increases the proportion of water not bonded. The unbonded molecules are in equilibrium with those of the clusters, the average life of which is about 10^{-11} s (Frank and Wen, 1957).

By reason of the three-dimensional nature of the lattice of molecules associated by hydrogen bonds, water is the most strongly 'structured' solvent. In other protonic-type solvents where the hydrogen bonds cause intermolecular associations, these latter generally form chains (in the case of alcohols) or rings (in the case of hydrofluoric acid : hexamolecular rings) and the degree of organization is much less pronounced than in water. The stability of the intermolecular associations itself varies between different solvents: the strongest hydrogen bonds are those in hydrofluoric acid, then come those in water, then in ammonia; on the other hand, they are so weak in hydrochloric acid and in sulphuric acid in the liquid form that these solvents are practically monomers, like non-protonic type solvents where cohesion is entirely due to the van der Waals forces.

The mixing of a highly structured solvent with another necessarily brings with it modifications to the structure; this occurs in a quite special manner in the case of hydro-organic mixtures. For example, when we add increasing quantities of alcohol to water, the alcohol molecules are at first incorporated in the system of water molecules, causing an enforcement of structure. Further, a high proportion of alcohol causes an effect in the opposite direction, appearing for a smaller proportion of alcohol as its molecule is more different from that of H_2O (Franks and Ives, 1966). Another example is that of mixtures of N-methylacetamide (NMA) and its isomer N,N-dimethylformamide (DMF). In pure NMA, the molecules form linear polymeric associations by hydrogen bonds with the two methyl groups in the 'trans' position); in pure DMF on the other hand, no molecular association is formed. The addition of a small proportion of DMF to the NMA then suffices to cause a progressive rupture of the molecular chains of the NMA, and therefore a structure-breaking effect (Dawson and Wharton, 1960).

2. Ionized solvents

These are no longer constituted from molecules, but from cations and anions, exerting electrostatic interactions on each other (Coulomb forces) which maintain the cohesion of the liquid. In contrast to what takes place in the solid state (ionic crystal), ions have a high mobility which gives this type of liquid a very high electrical conductivity (conductivity by ion electromigration).

The essential representatives of this class of ionized solvent are molten salts (although some of these are practically unionized and so belong to the family of molecular solvents; for example, covalent halides such as $SbCl_3$, $AsBr_3$, $HgBr_2$, which approach $POCl_3$, $SeOCl_2$, BrF_3, etc.).

The liquid state of ionic salts appears only at a relatively high temperature. Besides this, since the thermal stability of many molten inorganic salts is very high and their volatility is small, up to very high temperatures, these are the solvents suitable for high-temperature chemistry. Thus, sodium chloride melts at 808 °C and boils (under atmospheric pressure) at 1465 °C, sodium fluoride melts at 995 °C and boils at 1704 °C; other sodium salts melt at widely differing temperatures – $NaNO_3$ at 310 °C, NaOH at 318 °C, Na_2CO_3 at 854 °C, Na_2SO_4 at 884 °C, Na_2SiO_3 at 1089 °C – and remain liquid without appreciable decomposition over a fairly wide range of temperature.

The ionized salts liquid at the lowest temperatures are, notably, potassium thiocyanate (m.p. 177 °C) and ammonium thiocyanate (m.p. 88 °C), and especially various alkyl- or aryl-ammonium salts: ethylpyridinium bromide (m.p. 114 °C), ethylammonium chloride (108 °C), tetra-amylammonium thiocyanate (50.5 °C), etc.

So that it can be operated in an ionized liquid medium at as low a temperature as possible, mixtures of salts forming eutectics with melting points well below those of the pure constituents are often used. As an example, Table II gives the composition and melting point of some of the main eutectics used in practice.

As in the case of molecular solvents, a knowledge of the structure of ionized liquids plays an important part today in understanding their properties as solvents. Here again, experimental measurements (by X-rays in particular) have given proof that the ionic liquid is not entirely disordered, but retains an organization recalling that of the crystal lattice. Several hypotheses have been put forward to account for the differences between the liquid state and the solid state: existence of vacancies, that is unoccupied ionic sites in the crystal lattice (Frenkel, 1935; Kirkwood, 1935-1936; Bresler, 1939), existence of interstitial cavities (not corresponding to the crystal lattice) (Altar, 1937; Eürth, 1941), existence of dislocations in the crystal lattice, isolating 'icebergs' of monocrystalline structure (Stewart, 1927; Rothstein, 1955; Mott and Gurney, 1957), expansion leading to

TABLE II

Principal eutectic mixtures used as ionized solvents

Nature and melting points of constituents	Relative proportions of constituents (in moles %)	Melting point of eutectic (in °C)
$AgNO_3$(210 °C) : $TlNO_3$(206 °C)	48 : 52	82.2
$AlCl_3$: KCl(772 °C) : NaCl(808 °C)	60 : 14 : 26	94
KNO_3(337 °C) : $LiNO_3$(254 °C) : $NaNO_3$(310 °C)	53 : 30 : 17	120
KNO_3 : $LiNO_3$	59 : 41	133.5
KOH(360 °C) : NaOH(318 °C)	50.6 : 49.4	170
KBr(734 °C) : LiBr(547 °C)	40 : 60	348
KCl(772 °C) : LiCl(610 °C) : NaCl(808 °C)	24 : 43 : 33	357
BeF_2(797 °C) : NaF(995 °C)	60 : 40	360
KCl : LiCl	41.5 : 58.5	361
K_2CO_3(896 °C) : Li_2CO_3(735 °C) : Na_2CO_3(854 °C)	25 : 43.5 : 31.5	397
KF(856 °C) : LiF(845 °C) : NaF(995 °C)	42 : 46.5 : 11.5	454
LiF : NaF	50 : 50	484
K_2SO_4(1074 °C) : Li_2SO_4(859 °C) : Na_2SO_4(884 °C)	8.5 : 78 : 13.5	512
K_2SO_4 : Li_2SO_4	28.4 : 71.6	535

appearance of a free volume as in a compressed gas (Eyring, 1937); the most recent theories are associations of several of these hypotheses: vacancies in crystals, dislocations and free volume (Eyring *et al.*, 1958; Bloom and Bockris, 1964).

From a more chemical point of view, the constituent ions of an ionized solvent can be of several types. A medium such as molten sodium chloride is typically composed of monoatomic ions, simple and unique within their category (cations or anions), Na^+ and Cl^-. But other ionized media include also complex polyatomic ions.* Thus, in nitrates, carbonates, and sulphates, the anions are oxo-complexes (complexes with one central atom and oxide anions O^{2-}); the equimolecular mixture of aluminium chloride and sodium chloride is composed of simple Na^+ cations and complex tetrachloroaluminate anions $AlCl_4^-$ (including all the aluminium and chlorine atoms); in the same way, liquid cryolite (mixture of AlF_3+3NaF) is composed of various ions: Na^+, AlF_6^{3-}, AlF_4^- and F^- (Frank and Foster, 1957). Still considering the structural angle, we can reserve a special place for

* The distinction between a complex ion and an equivalent grouping of monoatomic ions with just electrostatic interaction can depend on various criteria: partial covalent character of interatomic bonds, average duration of aggregate. At the experimental level we can regard an entity as complex if the constituent atoms appear to move together, remaining associated, under the effect of diffusion or electromigration. Spectral considerations also lead to a precise notion of a complex.

ionized liquids containing complex polymerized anions: mainly silicates, borates, phosphates (mixtures in various proportions of the oxides SiO_2, B_2O_3, P_2O_5 with other metallic oxides) (Bloom and Bockris, 1964).

A third class of solvents, structurally different from the two preceding ones, can still be considered: liquid metals, some of which can be used at fairly low temperatures (mercury, m.p. —38.8 °C, sodium, m.p. 97.9 °C, eutectic with 33.5% sodium and 66.5% potassium, m.p. —12.6 ° C).

We are dealing there with a region which at present remains unexplored comparatively to the two other types of solvent.

There is in reality no fixed dividing line between the different categories of solvent which have just been briefly described, since 'mixed' media are very frequent. Thus, there are molecular liquids which are good conductors of an electric current, which indicates that there is already substantial self-ionization of the molecules; these solvents are intermediate between true molecular solvents and completely ionized solvents.

Mixtures of solvents sometimes pass from one type to another: two molecular solvents when mixed can produce an ionized medium (examples: mixtures of $H_2SO_4+H_2O$, composed essentially of H_3O^+ and HSO_4^- ions, and excess of molecules, either H_2SO_4 or H_2O; mixtures of aluminium bromide, practically unionized in the pure state and of pyridine, containing Br^- and complex (pyridine, $AlBr_2)^+$ ions; etc.).

It is also necessary to note that mixtures of salts and water where water predominates are generally considered to be electrolytic solutions in a molecular solvent, but a molten hydrate – for example: $Na_2SO_4,10H_2O$, m.p. 31.5 °C; $CaCl_2,6H_2O$, m.p. 29.8 °C; $LiNO_3,3H_2O$, m.p. 29.9 °C – is considered to be an ionized solvent (hydrated metallic cation+hydrated anion); it is therefore possible to consider these two cases as media intermediate between molecular solvents and ionized solvents.

Lastly, there are liquid mixtures of metals and of molecular liquids or of molten salts, which can still be considered as intermediate between the category of metals (electronic conductors) and that of ionized liquids (ionic conductors) or molecular liquids (non-conductors).

Gutmann (1968) drew up a very representative diagram showing the classification of all types of solvent possible, by means of a triangular diagram, with its apices occupied by the three limiting types defined above. Media intermediate between two categories are represented by points on the side of the triangle; electrolytic solutions near the apex for molecular solvents and molten solvates near the apex for ionized solvents, for example. The points inside the triangle then correspond to media which are intermediate between the three classes; this region is at present unexplored.

I.2. Solvation of Substances within a Molecular Solvent

In order to carry out chemical reactions inside a solvent, it is first necessary to be able to dissolve the substances which are to be the reagents. This dissolving process – that is the dispersion of the solute, either in the molecular state or in the ionic state, amongst the molecules of the solvent – brings several phenomena into play, some dependent on the substance and its state, others dependent on the solvent and its characteristics. It is necessary in fact for the molecules (or the ions) of the pure substance to be dissolved to detach themselves one from the other (except when they are in the gaseous state), then to insert themselves amongst the molecules of the solvent. This insertion is accompanied by the setting-up of interactions between the units of the solute and those of the solvent, known as solvation phenomena.

After dissolving a substance, the solution contains molecules or ions which are 'solvated'. The state of these foreign molecules or ions integrated into the homogeneous medium formed by the solvent liquid is one of the most discussed and most studied problems of physical chemistry of solutions and has been so for some years. We will describe it only very briefly with regard to its structural aspect; we will return in the last chapter, with more details as to its thermodynamic aspect, because of the link between this latter and questions of chemical reactivity.

Solvation of ions and molecules

The substances which give rise to solvated ions are called 'electrolytes'. Those which dissolve entirely in the state of molecules are 'non-electrolytes'. Amongst the substances which dissolve in the ionic state (anions and cations), we distinguish between 'ionophores', where the ions are already present in the pure substance (ionic crystals), and 'ionogens', where the ions appear on dissolving the molecular substance as a result of an action of the solvent known as 'solvolysis' (that is decomposition by action of the solvent).

1. To illustrate firstly the dissolution mechanism, let us consider that of an ionic crystal. The disaggregation of the crystal necessitates the removal of the cohesive forces existing inside it (electrostatic interactions between the constituent ions): each of the ions of the surface is subjected to a surrounding action by the solvent molecules, due to the interactions it can exert on them; then the ions so surrounded become detached from the crystal and spread out into the liquid. This latter phase of the dissolving process is facilitated by a high value of the dielectric constant of the solvent; in fact the greater this is, the more the electrostatic forces between the ions are weakened, enabling them to be dispersed within the solvent. This explains why liquids of high dielectric constant are the best solvents for ionic compounds.

Water is from this point of view in a favoured position, since its dielectric constant is close to 80; few other solvents have a higher dielectric constant. On the other hand, solvents of low dielectric constant and with only slightly polar molecules do not dissolve ionic compounds very much.

2. Overall, a solvated species can be defined as the entity formed by the foreign substance dissolved (ion or molecule of solute) and by the solvent molecules surrounding it and associated with it as a consequence of solvation interactions.

Let us consider in greater detail the case of a solvated ion. Amongst the solvating molecules around it, we distinguish those in contact (the most strongly associated), forming the 'primary solvation shell', and those which still interact, more weakly, from beyond this first layer, forming the 'secondary solvation shell'. The total number of molecules thus associated with the solvated ion is the 'solvation number'. Various experimental methods have been applied in order to determine values of this, notably for ions in aqueous solution (hydration number) – for example, electrodialysis experiments enabling us to determine the quantity of solvent having accompanied the ions on crossing a separating membrane. For the Na^+ ion alone, in water, figures as varied as 1, 2, 4 etc. (up to 66 and 71) have been put forward by different authors, showing the difficulty of obtaining significant results on this point (corresponding to the diffuse nature of the secondary solvation shell). In connection with the primary solvation shell only, the ideas and results of crystallography have been put to good use and the 'coordination number' of the ion then plays a certain part.

3. Another question is the nature of the strength of the interactions between solute and solvent. They may be of different sorts: van der Waals forces, electrostatic interactions ion-dipole or dipole-dipole (depending on whether the material solvated is an ion or a polar molecule, and when the solvent itself is dipolar), hydrogen bonds (when the solvent and the solute are respectively donor and acceptor of this type of bond) and then coordination bonds (complex formation between the solute and the solvent molecules).

The solvation of an anion in the common dipolar solvents calls into play essentially electrostatic interactions, with the positive end of the nearest dipole molecules of the solvent, with in addition, in the case of protonic solvents, hydrogen bonds. Parker (1962) distinguished, for solvation of anions, two categories of solvent: (a) protonic solvents, plentiful providers of hydrogen bonds – water, alcohols, ammonia, formamide and acetamide, etc. – in which the anions are solvated the most (especially in water), to an extent varying inversely with their size; (b) aprotonic solvents, very weak providers of hydrogen bonds – acetonitrile, acetone,

nitromethane, etc. – in which the anions are weakly solvated and, opposite to the previous solvents, the effect is less the smaller they are; solvation in fact depends on the polarizability of the anion, which increases with size. For example, the order of increasing solvation force of halide anions in solvents of the first category is $I^- < Br^- < Cl^- < F^-$; it is in the reverse order in solvents of the second category.

Cation solvation, in the same dipolar solvents, brings into play electrostatic interactions with the negative end of the nearest molecule-dipoles of the solvent, or in certain cases of the coordinating bonds, in the formation of which the surrounding solvent molecules play the role of donors of electron pairs. In this case, the greatest solvation is obtained with solvents where the molecules contain atoms with free doublets, nitrogen or oxygen, if the negative charge is well localized (water, ammonia, dimethylsulphoxide, dimethylformamide and dimethylacetamide, substances with phosphoryl groups, etc); but solvation is weakened if the negative charge is dispersed or masked by substitution groups of the solvent molecule (acetonitrile, nitromethane, tert-butanol) (cf. Price, 1966).

4. Another point of view is to consider the influence of the solute on the structure of the solvent. It can be agreed in a general way that the enveloping of the solute by solvent molecules takes place in accordance with a configuration corresponding to the smallest possible modification of

this latter. In support of this hypothesis we can quote the discovery that, for a series of monoatomic ions in dilute aqueous solution, the coordination number with near-by water molecules remains equal to the coordination number of the water molecules between those of the clusters, say four.

It is seen, however, that certain solutes have a structure-making ability on a molecular solvent, whilst others have a structure-breaking ability. We have seen above that alcohol in dilute aqueous solution enforces the structure of water; the same effect takes place with small ions which enter easily into the chains of water molecules (Li^+, F^-, OH^-), joining later to them by means of hydrogen bonds (F^-, OH^-), or again with large organic ions (quaternary ammoniums R_4N^+) by virtue of the hydrophobic nature of their alkyl or aryl groups which maintain the surrounding water very much associated. On the contrary, other ions have a structure-breaking effect on water (for example ClO_4^-, I^-, SCN^-, BF_4^-, $FeCl_4^-$, Ag^+).

Another conception, developed especially for ions in aqueous solution (Samoilov, 1957), is the consideration of the effect of the solute on the translatory movements due to thermal agitation of the solvent molecules situated in its vicinity. For the molecules of a pure liquid these movements are defined in the following manner: each molecule occupies, for a certain time, a site in the structure (ordered at short distance), in which it vibrates about an equilibrium position; when it possesses sufficient energy (of activation) to pass through the potential barrier, it jumps from this equilibrium site to another free site (a vacancy in the structure) near by, thus making a translatory movement. When an ion is introduced into the medium, the results of its interactions with the near-by solvent molecules is that these latter undergo a change in their mean time of remaining in one equilibrium site. All the molecules around the ion undergoing this perturbation are by definition those which take part in the solvation. The more strongly the solvent molecules are bound to the ion the longer the time of remaining; in the limit it becomes infinite for a permanent combination. On the other hand lack of modification corresponds to a lack of solvation. But the effect of certain ions is to cause a reduction in mean time in one position, which shows itself as a reduction in viscosity since the molecules have become more mobile. This phenomenon is known as 'negative solvation'. It is produced by large ions with small charges (K^+, Cs^+, I^-) in the most strongly structured solvents (water, polyalcohols), leading to the idea that it is due to the fact that the ion-molecule interactions of solvent which are set up are too weak to compensate for the structure-breaking effect of the ion.

SOLVATION OF THE ELECTRON IN CERTAIN MOLECULAR SOLVENTS

One of the most remarkable properties that have appeared in connection with the use of non-aqueous solvents is the solubility of metals in non-metallic solvents such as ammonia, aliphatic amines, hexamethylphosphorotriamide (HMPA).

Alkali metals are extremely soluble in liquid ammonia for example – the solvent in which solutions of metals have been most studied, starting from the first investigations of Weyl (1863) – ; their solubility exceeds 10 moles of metal per kilogram of ammonia, which is considerably greater than the solubility of sodium chloride in water, for example. The alkaline earth metals are also soluble, but to a smaller extent than the alkali metals; their solution enables ammoniates $M(NH_3)_6$ to be crystallized, analogous to ammoniates or hydrates of crystallized salts.

All these metals show the same blue colour when in dilute solution in ammonia, which corresponds to a wide luminous absorption band in the region of 7000 cm^{-1} due to a $s \rightarrow p$ electronic transition. The solutions are paramagnetic; study of the molar susceptibility and of the paramagnetic resonance spectra have led to evidence of the existence of free electrons in these solutions, located in cavities inside the liquid, with radii of 3 Å. The molar conductivity of these solutions, five or ten times that of aqueous saline solutions, also leads us to suppose the existence of electronic conduction, not ionic conduction as in electrolytic solutions. It has therefore been suggested (Kraus, 1921) that dissolved metals undergo in dilute solution a dissociation leading on one hand to solvated metallic cations and on the other hand to free electrons. These free electrons may be considered as solvated by the ammonia molecules surrounding them.*

Naturally, such solutions can exist only in solvents where the molecules are not going to be destroyed by reduction; this is why their numbers are very limited. In reality, even solutions of alkali metals and alkaline earth metals in ammonia or similar solvents are not stable, and decompose in the long term; in ammonia, for example, hydrogen and the alkali or alkaline earth amide are produced, in consequence of the reduction of a hydrogen ion from one solvent molecule. The presence of a catalyst (iron or platinum, finely divided) can accelerate this decomposition.

I.3. Ionic Associations ('Ion-pairs') in Molecular Solvents of Low Dielectric Constant

Since the time of the Arrhenius-Ostwald theory on electrolytic dissociation, it has been customary to distinguish in aqueous solution between 'strong electrolytes' – substances which are dissolved and completely dissociated into ions which can be regarded as independent one from the other (free ions) – and 'weak electrolytes' – dissolved substances which are not completely dissociated, one part being ionized and the other remaining as molecules.

* In concentrated solution, associations M^+e^- (M not dissociated) M_2 and Me^- (say M^-) have been assumed. Lastly, very concentrated solutions have structures tending towards those of liquid metals.

The state of total dissociation is shown by verification of certain laws, in particular the law of conductivity – observed empirically (Kohlrausch, 1900) before being deduced from the theory of Debye and Hückel (Onsager, 1926):

$$\Lambda = \Lambda_0 - kc^{1/2} \tag{I-1}$$

(Λ=equivalent conductivity, Λ_0=limiting equivalent conductivity at infinite dilution; c=concentration of electrolyte*; k=coefficient varying with the charge of the ions of the electrolyte). Divergences from this law are interpreted in terms of incomplete dissociation.

In aqueous solution, weak electrolytes are found essentially amongst ionogenic substances, those of which hydrolysis is not complete. Ionophoric substances behave – at least in not too concentrated solution – like strong electrolytes.

But it is found that an ionophoric substance, which behaves like a strong electrolyte in water, can, in another molecular solvent of lower dielectric constant, exhibit divergences from the conductivity law (I-1) similar to those observed for weak electrolytes. There can be no

* In aqueous solution the concentration is generally expressed in mol l^{-1} (molarity, M); in the case of solvents other than water this unit is often replaced by that of molality: the number of moles of solute per kilogram of solvent (mol kg^{-1}). In the two cases the concentration will be represented by the formula of the solute between square brackets: [ ]. We can also use the mole fraction, a dimensionless number, in place of the concentration.

question of incomplete ionization of the substance, which is already ionic in the pure state, so in order to explain this phenomenon the only possibility is an association of a greater or lesser part of the ions inside the solvent: aggregations of ions of opposite sign naturally reduce the number of charges able to take part in conductivity, hence the divergencies from Kohlrausch's law. Uncharged aggregates no longer take part either in the conductivity of in the 'ionic strength'* of the solution, and behave like molecular solutes.

Two associated ions of opposite charge constitute an 'ion-pair'. This type of association by pairs preponderates in most molecular solutions of low dielectric constant. Nevertheless, in apolar solvents of very low dielectric constant (benzene for example) it has been shown that there are triplets and quadruplets of ions (in small ratio to the numbers of ion-pairs).

Ionic association equilibrium

In a solution, ion-pairs and free ions are to be found in equilibrium. For example we can put:

$$A^+ + B^- \rightleftharpoons \text{ion pair } A^+B^- \qquad \text{(I-2)}$$

and we can characterize this equilibrium by an association constant:

* The ionic strength of a solution is defined by the formula $I_c = \frac{1}{2} \Sigma_i(z_i^2 c_i)$, where c_i is the concentration of each of the free ions, of charge z_i, present in the solution.

$$K_{ass} = \frac{[A^+B^-]}{[A^+][B^-]}. \quad \text{(I-3)}$$

Table III gives by way of example some experimental values of log K_{ass} in various solvents. It is seen that these values become very large in solvents where the dielectric constants are the least.

The value of K_{ass} enables us to calculate the degree of association α of the ions of an electrolyte of which the total concentration is c. For a 1-1 electrolyte, alone in solution, we have in particular the following correspondence between the values of α and those of the product $K_{ass}c$:

α %	:	50	90	99	99.9
$K_{ass}c$	:	2	90	10^4	10^6

Thus, in a 0.1 M potassium chloride solution in acetic acid (ε=6.2; logK_{ass}=6.9), about 99.9% of this salt will be in the form of ion-pairs K^+Cl^- and only 0.1% in the state of free K^+ and Cl^- ions (the ionic strength is therefore approximately 10^{-4} M). On the other hand, at the same concentration in methanol (ε=32.6; logK_{ass}= =1.15), the association rate of 0.1 M potassium chloride is only 32%, and 68% of the salt is in the state of free ions.

In benzene, 1,4-dioxane and chloroform there are practically no free ions; the ion-pairs always preponderate. For solvents of dielectric constants ranging from 5 to 10 approximately, we can conclude that the dissociated

TABLE III

Experimental values of log K_{ass} of some salts in various molecular solvents of low dielectric constant at ordinary temperature (K_{ass} in mol^{-1} l)

		Salts							
		Potassium chloride	Tetraethylammonium chloride	Pyridinium perchlorate	Sodium perchlorate	Tetrabutylammonium perchlorate	Tetrabutylammonium picrate	Tetraethylammonium picrate	Tributylammonium picrate
Solvents:	ε								
Ethanolamine	37.7	1.2							
Acetonitrile	36.2		1.9				1.15		
Nitrobenzene	34.8		1.9				1.1		3.7
Methanol	32.6	1.15					1.4		1.6
Ethanol	24.3	2.0							
Ammonia (at —33 °C)	22.4	3.0							
Acetone	20.7				3.65	2.0	1.65		
Methylethylketone	18.4		3.4					3.0	
Sulphur dioxide (at 0 °C)	15.4	4.1							
Phosphorus oxychloride	13.9		3.15						

Triethyl phosphate (at 40 °C)	12.9					2.9	2.95	
Pyridine	12.3					2.9		
Diethyl ethylphosphonate	10.6					3.35	3.45	
1,2-Dichloro ethane	10.4		6.3		3.8	3.65		7.7
o-Dichlorobenzene	9.9					4.75		
Dimethyl phthalate	8.4						4.35	
Diethyl phthalate	7.5						4.8	
Tricresyl phosphate (at 40 °C)	6.9					5.0		5.95
Acetic acid	6.2	6.9	6.1	5.5		5.8		
Chlorobenzene	5.6					7.7		12.7
Dioctyl phthalate	5.1						6.05	
Trioctyl phosphate (at 40 °C)	4.7					5.2	6.05	
Anisol	4.3					8.95		
Benzene	2.3				17.55	16.75		
1.4-Dioxane	2.2				18.95			
Mixtures of dioxane and water	41.5	0.1						
	30.2	1.25						
	25.8	1.55						
	19.3	2.2						
	15.3	2.7						
	12.8	3.25						

Values quoted in *Chemical Reactions in Solvents and Melts*, by G. Charlot and B. Trémillon, pub. Pergamon Press, 1969 (translated from French).

fraction of a salt in a sufficiently concentrated solution remains practically negligible in comparison with the part present in the form of ion-pairs. For ε going from 10 to about 40, the proportion of free ions ceases to be negligible, but there still is formation of ion-pairs, except in very dilute solution. It is only for $\varepsilon > 40$ that we can say that it behaves like a strong electrolyte and we can neglect the formation of ion-pairs, at least in solutions which are not too concentrated.

Theory of ion-pair formation

In solutions of lowest dielectric constant, where ionic dissociation therefore remains very small, the ionic strength of the electrolytic solutions cannot reach high values. The conductivity of these solutions remains low also, even when the electrolyte is concentrated.

The essential factor controlling the phenomenon of ion-pair formation therefore appears to be the dielectric constant ε of the solvent. Electrostatic Coulomb forces between ions of opposite sign, their magnitude increasing inversely as ε, are responsible for this.

It is already known, since the Debye–Hückel theory (1923), that these electrostatic forces, at long distance, are the cause of the divergences between the activity and the concentration of free ions in solution, divergences which increase with the ionic strength. It was by completing this theory that Bjerrum (1926) produced the first justification of the existence of ion-pairs and gave the first theoretical treatment of the subject.

In the treatment of Debye and Hückel, it is postulated that the energy of electrostatic interaction of the ions remains small compared with their mean kinetic energy of thermal agitation. But since the electrostatic force of attraction between two ions of opposite charge varies inversely as the square of the distance d between the centres of the two ions, Bjerrum took an interest in the region of interactions at short distances, where the Debye–Hückel treatment is no longer valid. He was thus able to show that there exists a critical distance d_c such that the energy of electrostatic interaction (work of separation of the two opposite charges to remove them from distance d_c to infinity) is equal to the mean kinetic energy of removal, that is four times the mean kinetic energy per degree of freedom ($\frac{1}{2}RN^{-1}T$). The d_c value is determined from the equality:

Coulomb attractive force
=centrifugal force due to thermal agitation.

Let:

$$\frac{z_1 z_2 e^2}{d_c^2 \varepsilon} = \frac{2RT}{N d_c} \tag{I-4}$$

whence:

$$d_c = \frac{z_1 z_2 N e^2}{2\varepsilon RT}. \tag{I-5}$$

In water at ordinary temperature (25 °C) this distance is about 3.5 Å for two ions of charges +1 and —1; it becomes 14 Å for two ions of charges +2 and —2.

Bjerrum then states that, if two ions of opposite sign are separated by a distance less than the critical distance d_c, they cannot separate by thermal agitation, and therefore form an ion-pair, contributing nothing either to the ionic strength of the solution or to the conductivity. On the other hand, ions at a distance $d > d_c$ are considered as free and justifying the theory of Debye and Hückel. According to this theory, formation of ion-pairs is naturally possible only if the sum of the radii of the associated anion and cation, $a = r_a + r_c$, known as 'distance of closest approach' of the two ions, is less than d_c; if $a > d_c$, all the ions remain free.

Further, the curve of probability of the presence of another ion of opposite charge at a certain distance d (varying between a and infinity) from the centre of an ion – a curve with a minimum at $d = d_c$ – enables us to calculate the proportion of ions at a distance less than d_c, and hence associated, and the complementary proportion of ions which are at a distance greater than d_c and are free; hence the value of the association constant. The greater d_c, the greater the first proportion and the smaller the second, and in consequence the greater the association constant.

In a solvent such as water, with a high dielectric constant, the critical distance is generally less than the distance of closest approach of the ions (hydrated) and no ion-pairs are formed. It is only very small or highly charged ions which can exhibit a weak ionic association in concentrated solution. As an example, the experimental values of $\log K_{ass}$ (K_{ass}/mol^{-1} l) of some salts in aqueous solution at 20 °C are as follow (from Davies, 1962):

RbCl	—0.8	Na_2SO_4	0.7
NaOH	—0.7	K_2SO_4	1.0
$NaNO_3$	—0.6	$ZnSO_4$	2.3
KNO_3	—0.2	$Fe(ClO_4)_3$	1.15
$AgNO_3$	—0.2		

But in a solvent of low dielectric constant ($\varepsilon <$ about 40) the critical distance d_c, inversely proportional to ε, becomes greater that the closest approach distance for all electrolytes. As a result the proportion of associated ions cannot be neglected.

The theory developed by Bjerrum gives an expression from which the value of the association constant may be calculated:

$$K_{ass} = \frac{4\pi N}{1000}\left(\frac{z_1 z_2 N e^2}{\varepsilon RT}\right)^3 \int_2^b \frac{\exp y}{y^4}\,\mathrm{d}y = \\ = Q(b)\frac{4\pi N}{1000}\left(\frac{z_1 z_2 N e^2}{\varepsilon RT}\right)^3 \qquad \text{(I-6)}$$

where:

$$y = \frac{z_1 z_2 N e^2}{\varepsilon RT}\cdot\frac{1}{d} = 2\,\frac{d_c}{d}$$

and

$$b = \frac{z_1 z_2 N e^2}{\varepsilon R T} \cdot \frac{1}{a} = 2\frac{d_c}{a}. \quad \text{(I-7)}$$

The function $Q(b)$ (a transcendental exponential integral function) can be calculated numerically; there are tables of values in existence which can be used for calculation of K_{ass}.

Denison and Ramsey (1955), and then Gilkerson (1956) have given a simplified version of Bjerrum's treatment, leading to a simpler relation for the value of K_{ass}; the same expression was adopted and derived theoretically by Fuoss (1958), who approached the problem of ionic associations at the theoretical level:

$$K_{ass} = \frac{4\pi N a^3}{3000} \exp b\,. \quad \text{(I-8)}$$

Let:

$$\log K_{ass} = \text{constant} + \frac{0.43 z_1 z_2 N e^2}{aRT} \frac{1}{\varepsilon}. \quad \text{(I-9)}$$

According to this expression, $\log K_{ass}$ must vary inversely with the dielectric constant of the solvent. Fuoss' theory differs from that of Bjerrum in particular in that it is only the ions at the distance of closest approach that are considered to be associated and so forming an ion-pair.

Under any circumstances, the theory of ionic associations can only be approximate, for various reasons: real ions are not all spherical, with their charges symmetrically distributed, as theory postulates (polarity also affects the result); the dielectric constant varies in the neighbourhood of the ions, falsifying the calculation of the electrostatic energy uniting the ion-pairs; on pairing it is certain that modifications to the state of solvation of the ions are produced, affecting the energy; finally, pairing can be accentuated by intervention of other types of interaction between the ions, especially hydrogen bonds (notably in the case of amine salts).

It therefore appears almost useless to seek a very close experimental verification of the theory, in particular with a view to deciding in favour of Bjerrum's or Fuoss' treatment. Attempts at verification of the law of linear variation of $\log K_{ass}$ with the reciprocal of the dielectric constant, using mixtures of solvents giving a progressive variation of ε, are sometimes crowned with success, but very frequently are not; this is not a confirmation of Bjerrum's law, since it is also necessary to allow for the possible variations in radius of the solvated ions, and hence of the distance of closest approach, when the nature and composition of the solvent vary.

To summarize, all dissolved compounds, even when completely ionized, behave, in molecular solvents of low dielectric constant, like weak electrolytes. A fortiori, compounds which are not completely ionized are still less dissociated than simple ion-pairs. This fact gives in principle a means of distinguishing these latter from complexes, but the distinction ceases to be possible for the weakest complexes, unless the behaviour of the limiting case of the ion-pair is determined with great precision.

In any case, ionic association or complex formation have the same effect on the activity of free ions and therefore modify the reactivity of these latter in the same manner. This is why the phenomenon of ionic association, which is inevitable in solvents of low dielectric constant, enters into the arguments used in the case of these solvents, and plays a part which cannot be neglected in the treatment of chemical properties. We will see this in Chapter II on the subject of the properties of acids and bases.

I.4. State of Substances Dissolved in an Ionized Solvent (Molten Salt)

Their ionized structure and high temperature range of application do not prevent molten salts from behaving as excellent solvents for all types of thermostable substance: molecular species (gaseous, liquid or solid at the temperature of the solution, organic or inorganic), ionophoric or ionogenic substances, and in certain cases metals, all as in molecular solvents.

In the same way, dissolution calls into being interactions between the solute and the species of ionized solvent which surround it, interactions which correspond in a similar manner to a 'solvated' state of the foreign matter introduced. These interactions are generally speaking of the same nature as those set up inside other solvents: electrostatic forces, van der Waals forces, coordination bonds, even in certain cases hydrogen bonds. But, having regard to the ionic nature of the solvent, one type of interaction becomes of prime importance in this case: the Coulomb forces of attraction and repulsion between electrically charged entities.

Ionic solutes

A simple structural model has been suggested for solute ions where the only interaction with ions of solvent (assumed to be completely ionized) is electrostatic in nature. In this model a molten salt is regarded as a form of crystal lattice, made up of two distinct sub-lattices each penetrating the other: on one hand the cation sub-lattice, and on the other hand the anion sub-lattice. The intense Coulomb electrostatic forces operating lead to an alternation of electric charges; a cation must be surrounded by anions and an anion must be surrounded by cations. Random distribution therefore only occurs for cations in the sites of the cation sub-lattice, and on the other hand, for anions in the anion sub-lattice.

Another salt having an ion in common with the solvent, the anion for example, can be dissolved in a molten salt. The solute cations will then distribute themselves over the sites of the cation sub-lattice only. When they have an electric charge greater than that of the solvent cations, their introduction into the cation sub-lattice in the place of a solvent cation is accompanied by the appearance of empty sites (vacancies) in this sub-lattice, in such a manner as to give a neutral distribution of electric charges (Førland, 1957).

In the case where the anions of the solute salt are in addition different from the anions of the solvent, the same phenomenon is repeated for them as for the cations: they will distribute themselves over the sites of the anion sub-lattice, ultimately forming vacancies in this sub-lattice when their charge is greater than that of the solvent anions.

This structural model can remain valid in its essentials for the case of interactions between solute ions and ions of opposite sign to the solvent, when these interactions are more complex than merely electrostatic forces. Thus it has been shown that molten zinc chloride forms a polymerized lattice (lamellary) of Zn^{2+} and Cl^- ions, associated by complexation (chloride bridges), and hence dissolved Ni^{2+} cations are found to be inserted in this lattice in place of Zn^{2+} cations.

Introduction of a simple solute ion into a molten salt can also lead to formation of new complex entities with the solvent ions, very different from these latter (causing in consequence a degree of disorder in the structure in their neighbourhood, which is often substantial). The state of the solute is then similar to that of a substance which is solvated, by coordination or by hydrogen bond, in a molecular solution. For example, in molten alkali chlorides, it has been shown that the dissolved Ni^{2+} ions are strongly associated with the Cl^- ions surrounding them, forming a chloro-complex $NiCl_4^{2-}$ (to be compared with the hydrated $Ni(H_2O)_6^{2+}$ ion in aqueous solution); in molten alkali hydroxides, the Zn^{2+} dissolved ion is associated with the neighbouring OH^- ions to form the hydroxo-complex $Zn(OH)_4^{2-}$, etc. It can be said that the $NiCl_4^{2-}$ group in molten chlorides, and the $Zn(OH)_4^{2-}$ group in molten hydroxides are solvated Ni^{2+} or Zn^{2+} ions (solvated by the solvent anions).

The preceding structural ionic model is ultimately applied to these complex groups. Let us consider for example the case of the Cu^+ ion dissolved in molten mixtures of aluminium chloride and sodium chloride. In a mixture where the molar proportion of $AlCl_3$ is slightly greater than that of NaCl, the solvent consists of a sub-lattice of Na^+ cations and a sub-lattice of $AlCl_4^-$ anions, including a small proportion of $Al_2Cl_7^-$ anions (corresponding to $AlCl_3$ in excess in comparison with $Na^+AlCl_4^-$); in this mixture, the Cu^+ ion remains free and inserts itself in the sub-lattice of the Na^+ cations. On the other hand, in a mixture where the mole fraction of $AlCl_3$ is less than that of NaCl, the solvent still consists of the same sub-lattices of Na^+ cations and $AlCl_4^-$ anions, this time with the second one containing a proportion of free Cl^- anions (corresponding to NaCl in excess in comparison with $Na^+AlCl_4^-$); in this mixture, the Cu(I) ion forms complex species $CuCl_3^-$ with the Cl^-ions, which then insert themselves into the anion sub-lattice, and no longer in the cation sub-lattice as in the previous case.

Molecular solutes

For molecular solutes, the above considerations based on electrostatic effects naturally no longer apply, except where the solute becomes ionic by reasons of solvation by certain solvent ions; for example: chlorine or iodine, which dissolve in molten chlorides or iodides in the state of Cl^- or I_2Cl^- anions; hydrofluoric acid, which dissolves in molten fluorides in the form HF_2^-; $TiCl_4$, which dissolves in molten chlorides in the form $TiCl_6^{2-}$, etc.

A solute can remain in the molecular state by inserting itself in the empty, electrically neutral spaces (holes, vacancies, or free volume) of the ionic lattice of the solvent; the interactions between the solute and neighbouring ions are of the type ion-dipole or van der Waals, as for example: solutions of rare gases, solutions of sulphur, probably in the form of S_2 or S_4 molecules, paramagnetic and blue, in molten halides.

Solutions of metals. 'Solvation' of the electron

Like certain molecular solvents, molten salts are, in some cases, good solvents of metals. Bunsen and Kirchhoff noted this in 1861 in connection with rubidium and caesium (obtained by electrolysis) in their respective molten chlorides. With the exception of oxidation-reduction reactions between the metal and the molten salt, metals are soluble only in their own salts, more especially in halides: alkali metals in alkali halides, cadmium in cadmium halides, bismuth in bismuth halides, etc. The solutions are frequently deeply coloured: dark blue for sodium, intense red for cadmium. Alkali metals are also soluble in their molten hydroxides – giving dark green solutions – or in sulphides, etc.

A great number of metal-salt systems where the salt is molten have been studied experimentally at the present time, particularly from the point of view of phase equilibrium (cf. Bredig, 1964). The solubility of metals varies greatly, depending on the system and on the temperature. It can be high, sometimes reaching complete miscibility (as an example: caesium in caesium halides, at all temperatures at which the system is liquid; potassium in KCl, above 790 °C; bismuth in $BiCl_3$, above 780 °C, etc.) When miscibility is not total, two immiscible liquid phases – molten salt saturated with metal and metal saturated with molten salt – can appear.

Bredig has shown that two categories of solution can be distinguished, depending on the state of the metal dissolved in the molten salt. The first category is that where the solutions retain the metallic character of the solute. Solutions of alkali metals in their halides are a typical example of this; they recall solutions of these metals in liquid ammonia. The laws of conductivity of these solutions lead to deducing the existence of free mobile electrons occupying, according to the previous structural model, sites in the anion sub-lattice of the molten salt (Cubicciotti, 1952; Bronstein and Bredig, 1958).

In all other cases, corresponding to the second category of solutions of metals in their molten salts, it is deduced that the atoms of dissolved metal exchange the electrons which they provide with the metallic cations constituting one of the ionic groups present in the solvent, or that they combine with these cations, thus leading to the formation of a metallic cation in an intermediate oxidation state (sub-halide). For example, bismuth dissolved in molten $BiCl_3$ forms, by exchange between 2Bi and Bi^{3+}, Bi^+ cations (corresponding to bismuth monochloride); lanthanum dissolved in $LaCl_3$ forms La^{2+} cations in the same way. Mercury dissolves in mercuric chloride in the form of Hg_2^{2+} mercurous ions (corresponding to calomel Hg_2Cl_2); in the same manner, cadmium in molten $CdCl_2$ forms Cd_2^{2+} cations, resulting from the addition of an atom of cadmium and a Cd^{2+} cation (the distinction between Cd_2^{2+} and Cd^+, which both correspond to the same oxidation state of cadmium, is possible because the solution is diamagnetic, whilst Cd^+ would be paramagnetic) (cf. Corbett, 1964).

It is clear from the phase diagrams that various salts have appreciable reciprocal solubility in their liquid metals; this leads us to consider liquid metals as future solvents.

Ionic associations in molten salts

It frequently happens that a solute cation and anion form associated compounds when in solution in a molten salt.

A very stable association corresponds to the formation of a complex, similar to what takes place in a molecular solvent.

Theories based on the quasi-crystalline structural model (Flood *et al.*, 1953–1954, Blander and Braunstein, 1959–1961) have shown that ionic associations can be produced without calling into play any interactions other than Coulombic forces. It is then a question of a phenomenon of ion-pair formation, recalling that produced in molecular solvents of low dielectric constant. The result is an effect known as 'reciprocal Coulomb effect'.

If we dissolve a cation X^+ and an anion Y^- in an ionized molten salt A^+B^- used as solvent, and if X^+ and Y^- form (partially) an ion pair $\overset{+}{X}\overset{-}{Y}$, we can characterize the equilibrium of formation:

$$X^+ + Y^- \rightleftharpoons \overset{+}{X}\overset{-}{Y} \qquad \text{(I-10)}$$

by the association constant:

$$K_{ass} = \frac{[XY]}{[X^+]\,[Y^-]}. \qquad \text{(I-11)}$$

The theory developed by Blander (1959–1961, 1964) leads to an expression obeyed by K_{ass}:

$$\log\left(1 + \frac{K_{ass}}{Z}\right) = -\frac{\Delta G}{2.3RT} \qquad \text{(I-12)}$$

where Z is the coordination number of the central ion (metallic cation) and ΔG the free enthalpy change corresponding to the exchange, within the solvent, of an anion B^- in contact with X^+ for an anion Y^-, so forming an ion pair X^+Y^-. The following diagram illustrates the exchange process:

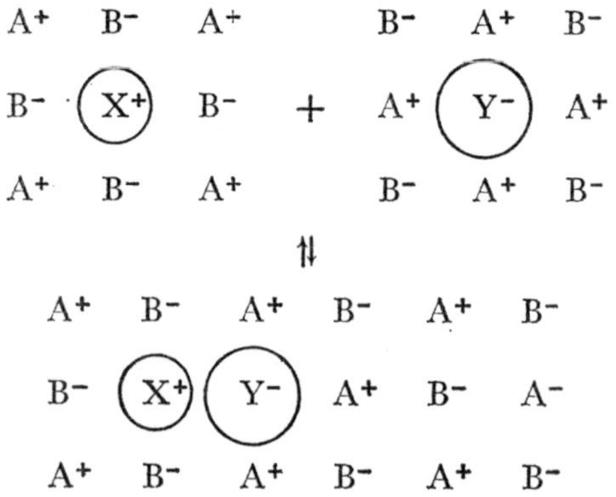

$$
\begin{array}{ccccccccc}
A^+ & B^- & A^+ & & & B^- & A^+ & B^- \\
B^- & (X^+) & B^- & + & A^+ & (Y^-) & A^+ & \\
A^+ & B^- & A^+ & & & B^- & A^+ & B^-
\end{array}
$$

$$\rightleftharpoons \qquad \text{(I-13)}$$

$$
\begin{array}{cccccc}
A^+ & B^- & A^+ & B^- & A^+ & B^- \\
B^- & (X^+) & (Y^-) & A^+ & B^- & A^- \\
A^+ & B^- & A^+ & B^- & A^+ & B^-
\end{array}
$$

In the case where the only interaction between the X^+ and Y^- ions consists of electrostatic Coulomb forces, and also for A^+ and B^- ions (considered as rigid spheres in contact), we can calculate ΔG from the energies of attraction of the simple ion-pairs X^+B^-, A^+Y^-, X^+Y^- and A^+B^-; these energies are inversely proportional to the distances d between the centres of the ions under consideration (sums of ionic radii: respectively d_{XB}, d_{AY}, d_{XY} and d_{AB}, for the different ion-pairs). ΔG is therefore of the form:

$$\Delta G = -k\left[\frac{1}{d_{XY}} + \frac{1}{d_{AB}} - \frac{1}{d_{XB}} - \frac{1}{d_{AY}}\right]. \qquad \text{(I-14)}$$

Its sign is determined by that of the term in brackets. We thus see mathematically, by considering relations (I-12) and (I-14), that two cases occur (the radii of the four ions concerned are designated by r_A, r_B, r_X and r_Y):

(a) if

$$\left.\begin{array}{l} r_X < r_A \quad \text{and} \quad r_Y < r_B \\ r_X > r_A \quad \text{and} \quad r_Y > r_B \end{array}\right\} \quad \Delta G < 0 \quad \text{and} \quad K_{ass} > 0 \tag{I-15}$$

(with "or" between the two conditions)

(b) but if

$$\left.\begin{array}{l} r_X < r_A \quad \text{and} \quad r_Y > r_B \\ r_X > r_A \quad \text{and} \quad r_Y < r_B \end{array}\right\} \quad \Delta G > 0 \quad \text{and} \quad K_{ass} < 0 . \tag{I-16}$$

(with "or" between the two conditions)

A negative value of the association constant has no physical meaning, implying only that the ion-pair XY cannot be produced.

In fact, if the cation radii r_X and r_A on the one hand, and the anion radii r_Y and r_B on the other hand, are respectively equal, we find $\Delta G=0$, and $K_{ass}=0$. This means that X^+ is randomly distributed over the cation sub-lattice and that Y^- is also randomly distributed over the anion sub-lattice. We cannot speak of association of X^+ and Y^-, since these ions show no particular tendency to remain in contact rather than in contact with solvent ions.

On the other hand, if the cation and/or anion radii differ amongst themselves, the distribution over the sub-lattice is not completely random. The reciprocal Coulomb effect favours (in accordance with I-14) contact between the smallest cation and the smallest anion, and of the largest cation and the largest anion. In consequence, if the solute cation X^+ and the solute anion Y^- are each smaller or larger at the same time than their homologue in the solvent, they will tend to associate ($K_{ass} > 0$), to a degree which increases as the difference in size becomes more marked. On the other hand, if one of the solute ions is larger than its homologue in the solvent and the other is smaller, they will tend to be less frequently in contact than in the ideal model, so explaining the negative value then obtained for K_{ass}.

The reciprocal Coulomb effect justifies in the same way the existence of higher associations of ions – triplets and quadruplets of solute ions, such as XY_2^-, XY_3^{2-} ..., X_2Y^+, etc. – which may be formed by the same mechanism. Relations, similar to (I-12), between their successive formation constants and the different energies called into play by each association, have been established (cf. Blander, 1964).

The influence of differences in respective sizes of solvent and solute ions is shown by the study of the variation of ionic association constants, as functions of the composition of a solvent composed of a mixture of two salts chemically similar but with different ion sizes; this is the case, for example, for the association of Ag^+ or Cd^{2+} ions with halide ions Cl^-, Br^-, I^-, in a molten mixture of potassium and sodium nitrates (Blander *et al.*, 1961–1962; cf. Blander, 1964). Within experimental error, the constants measured ($\log K_{ass}$) vary, between the values for each nitrate in a pure state, linearly with the mole fraction of one of the solvent constituents. The constants are highest in pure molten KNO_3, and lowest in pure $NaNO_3$; for example, for AgBr at 400 °C, $\log K_1 = 2.97$ and $\log K_2 = 2.57$ in molten KNO_3, whilst in molten $NaNO_3$, $\log K_1 = 2.80$ and $\log K_2 = 2.39$ (activities expressed by the mole fractions). The direction of variation is in accordance with that of the reciprocal Coulomb effect, since we deduce from the relationship (I-14) that $-\Delta G$ increases when we replace Na^+ by K^+ of greater radius (Ag^+ being smaller than Na^+ and K^+, Br^- being smaller than NO_3^-). Nevertheless, the theoretical calculation of the reciprocal Coulomb effect shows a variation which should be about twice as large as it is in practice; Blander attributes this divergence to the fact that Coulombic forces from long distances reduce the effect calculated by only allowing for interactions between ions in contact.

In any circumstances, it is evident that the reciprocal Coulomb effect in not on its own capable of explaining all the associations which are formed in molten salts, and to account for the observed values of association constants. More specific interactions, sometimes with greater effect, can simultaneously intervene.

As in the case of ion-pairs in molecular solvents of low dielectric constant, the distinction between weak complexes and purely ionic associations due to the reciprocal

Coulomb effect appears hardly possible by experimental means.

I.5. Presentation of Reactions in Solution

Solution chemistry is concerned with substances (dissolved or otherwise) which, placed in contact within a solvent, react amongst themselves, that is they undergo a transformation resulting in the formation of other substances. These latter frequently remain in solution, but it can happen that if the concentration reaches the level of solubility, 'saturation' takes place in the solution, and a substance appears either solid (precipitation) or liquid (separation) or gaseous (release of gas).

The chemical reaction corresponding to spontaneous evolution of the system composed of the mixture of reagents, takes place, with variation in the proportions of dissolved substances, in principle until equilibrium is reached. This is characterized by a relation between the activities of all the participants, constituting the law of mass action.* The value of the equilibrium constant enables us to calculate, having regard to the material balance, the values of the activities and hence the concentrations of the different participants in equilibrium when this latter is established. It is then possible to predict

* The activities of solid and liquid substances which saturate the solution may be taken as unity and are considered to be invariable so long as the solution remains saturated.

the chemical reactions which may be produced in a given solvent, on the assumption that equilibrium is reached sufficiently rapidly (if not, complementary kinetic conditions are necessary).

It would be incorrect to think that the solvent, whatever it may be, is just a bearer of the reactive species. On the contrary, it plays a part in the reactions and equilibria established inside it. It is just this reactivity of the solvent which is the cause of the limitations on the possibilities of chemical operations in the medium which it constitutes.

Recognition of the participation of the solvent in the different types of reaction and chemical equilibrium, its laws and the consequences flowing from them, are the subject of investigations into the chemical properties in non-aqueous solvents. We will devote the following chapters to examining them.

Depending on the methodology used for aqueous solutions, we continue to classify properties and chemical reactions in other solvents by:

(a) acid-base or protolysis phenomena;
(b) more general ion-exchange phenomena, into which complex formation phenomena enter;
(c) oxidation-reduction phenomena.

In the three following chapters we will consider theories giving a systematic treatment of each one of these three categories of phenomena in all solvents, stressing particularly the participation of the solvent in these phenomena.

NOTE CONCERNING THE SIMPLIFIED FORMULATION OF SOLVATED SPECIES FOR THE SYMBOLIC REPRESENTATION OF REACTIONS AND CHEMICAL EQUILIBRIA IN SOLUTION

Although dissolved chemical species are, in the case of molecular solvents, associated with solvent molecules, forming sometimes very strong combinations, we represent these solvated species just by the formula of the solute, not referring to the associated solvent material (that is by the formula for the part foreign to the solvent in the solvated material).

With regard to the reaction between a solvated species and another within a solvent liquid – forming for example a combination (itself solvated) – the structures of solvation are implied when this latter takes place, and in general there is produced, between the initial state and the final state, a variation in the total number of solvent molecules concerned in the different solvations. It results from this that reactions and chemical equilibria in solution should, in order to take account of these facts and show clearly the role of the solvent, be symbolized by use of chemical formulae in which the solvation of the species appears – for example, for a substance A in a molecular solvent S, in the form A,nS – and showing explicitly the solvent molecules liberated, or alternatively absorbed, by the changes in solvation during the reaction. Thus, for the formation of the combination AB, starting from A and B in solvent S:

$$\text{A}, n_1S + \text{B}, n_2S \rightleftharpoons \text{AB}, n_3S + (n_1 + n_2 - n_3)\, S\,. \tag{I-17}$$

Nevertheless, since the solvation numbers n cannot in general be specified with certainty, and so as not to overload the symbolic description of chemical reactions and equilibria, it has been agreed not to show the solvent molecules. The equilibrium (I-17) is therefore written more simply:

$$\text{A} + \text{B} \rightleftharpoons \text{AB} \tag{I-18}$$

as if the reaction were not taking place in a solvent. But this simplification should not make us forget that the reactivity of the substances thus shown in the formulae – in the same manner whatever S may be – contributed to the solvation and in consequence changes with the nature of the solvent.

To justify fully the simplified formulation, we must note that if the expression of the mass action law applied to a chemical system in the equilibrium state does in theory introduce the activity of the solvent S, this latter can be regarded as practically invariable and very close to unity when the solution is sufficiently diluted; it then disappears also from the expression of the mass action law, which only introduces the activities of substances A, B, and AB (solvated), which amounts to the application of this law to the system formulated as in (I-18).

For quite similar reasons, it is agreed to adopt the same simplification in writing the equations for species solvated in molten salts: the only part shown is that foreign to the solvent, omitting any reference to all the ions of the latter which are linked more or less energetically to the solute and whose activity remains practically invariable when the solution is sufficiently diluted.

For example, the Ni(II) ion dissolved in molten sodium chloride will be symbolized, to formulate a chemical reaction, by Ni^{2+} although the true substance is $NiCl_4^{2-}$; the Cl^- ions linked in the complex tetra-coordinated species form part of the solvation and will not be distinguished from the other Cl^- ions of the solvent. In the same way, $Zn(OH)_4^{2-}$ in molten alkali hydroxides will be shown simply as Zn^{2+}, etc.

This naturally amounts to attributing to the solute, contrary to the cases of molecular solvents, an electric charge different from the true over-all charge carried by the solvated material; but this is no inconvenience except when we are concerned with properties where the charge of the material is relevant (electromigration in particular).

CHAPTER II

ACIDITY AND PROTOLYSIS PHENOMENA

The concept of 'acidity' has for a very long time occupied a very special place in solution chemistry. By 1786, Guyton de Morveau had already written "to hold the definition of acids is to hold the key to chemistry".

The first idea of 'acids' and 'bases' goes back to the seventeenth century, but then consisted only of a classification into two groups of compounds capable of reacting between themselves to form 'salts'. Distinctive features of the two groups were recognized: for example, the taste of acids and their action on the colour of litmus solution, different from that of bases. In 1777, Lavoisier suggested the first definition base on considerations of composition; he regarded oxygen as the constituent element which led to the properties of an acid. But this idea had to be abandoned when Davy showed that hydrochloric acid was composed of hydrogen and chlorine only (1810). The definition most generally accepted for a long time after this was that of Liebig: acids are compounds containing hydrogen which can be replaced by a metal (1838).

The concepts of Arrhenius and Ostwald on electrolytic

dissociation in aqueous solution (1887) led to the first true theory of acids and bases. Arrhenius defined acids as compounds which, when dissolved in water, dissociated to form hydrogen ions (hydrated proton) and anions, whilst bases are compounds which dissociate into hydroxide anions and cations. A quantitative treatment of the reactions between acids and bases – leading to the combination of the hydrogen ions of the acid with the hydroxide anions of the base, to form molecules of water – based on the laws of chemical equilibria, then became possible. The idea of 'neutrality', the result of 'neutralization' of an acid by a base or vice versa was introduced.

As far as aqueous solutions are concerned, this theory of Arrhenius showed itself to be satisfactory and completely acceptable. But it appeared inadequate when interest started to be taken in solvents other than water, and when attempts were made to rationalize the properties of acids and bases in all solvents. Thus, ammonia gas dissolved in anhydrous acetone is able to react with acids, without bringing into play any hydroxide ions in the solution of ammonia in acetone, and any formation of water molecules during neutralization.

Independently, and at about the same time (1923), Brønsted and Lowry suggested a new definition of acids and bases independent of the nature of the solvent. This concept, more general than that of Arrhenius, can include this latter as a special case corresponding to aqueous solutions. Above all, the definition given by Brønsted and Lowry is far more rational, since it introduces the idea of an acceptor-donor couple (with occurrence of proton), an essential idea in the whole system of chemical reactions. The theory of Brønsted and Lowry is the most commonly accepted theory of acidity, and its application to the case of aqueous solutions is well known today.

Reminder of definitions of acids and bases according to the Brønsted-Lowry theory

Acids are compounds capable of playing the role of donor of a hydrogen atom, or proton, H^+ (protogenic compounds) and bases are compounds capable on the contrary of playing the role of acceptors of H^+ ions (protophilic compounds).

After taking a proton, a base is transformed into an acid; reciprocally, an acid losing a proton is transformed into a base. Hence, to every acid and every base there is a corresponding base and acid, known as the 'conjugate' of the initial body. The conjugate acid and base form an acid-base couple, a couple of donor and acceptor of an H^+ hydrogen ion, which can be symbolized by a relation of type:*

$$\underset{\text{(donor)}}{\text{acid}} = \underset{\text{(acceptor)}}{\text{base}} + H^+ . \qquad \text{(II-1)}$$

The definition of acids and bases assumes nothing as to

* The sign = shows here that a definition is being given.

their electric charge; they can be either molecular or ionic substances.

The same compound can at the same time play the parts of acid and of base; it is then said to be 'amphoteric'. Water is a typical example of an amphoteric body:

$$H_2O\,(\text{acid}) = OH^- + H^+$$

$$H_3O^+ = H_2O\,(\text{base}) + H^+\,.$$

The hydroxide anion OH^- is the conjugate base to water, whilst the hydronium cation H_3O^+ is the conjugate acid.

Like water, many other molecules capable of use as molecular solvents, exhibit this amphoteric character: alcohols, carboxylic acids, ammonia and primary or secondary amines, amides (not NN-disubstituted), etc.

Protolysis. – Protolysis is the general name for phenomena dealing with acids and bases.

An 'acid-base' reaction is a reaction exchange (or transfer) of a hydrogen ion between an acid (donor of H^+) and a base (acceptor of H^+) which are not conjugate. Let HA be the acid and B the base (without specifying their electric charge) which are brought together, each one belonging to a couple:

$$HA = A + H^+$$

$$B + H^+ = HB\,.$$

The exchange reaction is written:

$$HA + B \rightarrow A + HB. \tag{II-2}$$

When the reaction leads to a state of chemical equilibrium – which is most frequently the case, and is the only one we will consider in what follows – we replace the $\rightarrow$ sign by $\rightleftharpoons$. We are then dealing with an acid-base equilibrium: $HA+B \rightleftharpoons A+HB$. The mass action law leads to the expression for the equilibrium constant as a function of the activities of the participants:

$$K^0 = \frac{a_A \cdot a_{HB}}{a_{HA} \cdot a_B}. \tag{II-3}$$

Assuming that the activity coefficients (ratio of activity to concentration) are kept constant – reaction in dilute solution or at constant ionic strength – we can use an 'apparent' constant K, not very different from K° (if the activity coefficients do not greatly differ from unity):

$$K = \frac{[A]\,[HB]}{[HA]\,[B]}. \tag{II-4}$$

If the value of K is large, the product of the concentrations of the reagents HA and B brought into contact must in equilibrium be very small compared with that of the concentrations of the products A and HB; this means that equilibrium is shifted from left to right and that the acid-base reaction in this direction between HA and B is 'practically quantitative'. The contrary is true if the value of K is very small ($-\log K$ is large): the reaction between HA and B is 'practically nil' and the reaction which can take place is, on the contrary, that between A and HB.

Classification of solvents from the view-point of acidity-basicity

It is possible to distinguish several groups of solvents from the view-point of acidity-basicity:

1. A certain number of solvents exhibit simultaneously a 'protophilic' character and a 'protogenic' character. These are the protic solvents, amphoteric, known now as 'amphiprotic solvents'. Water is the main member of this class and is the model. Others are the alcohols (acid-base properties very similar to those of water), non-substituted amides (the protogenic feature corresponds to an H atom of the amide group), acids and in particular carboxylic acids (the protophilic character is conferred by the carbonyl group), liquid ammonia and the primary and secondary amines, and various solvents such as dimethylsulphoxide, acetonitrile, etc.

2. Another group of solvents consists of those which exhibit protophilic properties but not protogenic. The most common are ketones and ethers, such as acetone, 1,4-dioxane, tetrahydrofuran (protophilic because of the oxygen atom); some di-N-substituted amides such as

dimethylformamide, hexamethylphosphorotriamide; esters, propylene carbonate, sulpholane, etc.

We can in fact expect most of these solvents to have a protogenic character on theoretical grounds, due to one of the hydrogen atoms in their molecule, but so small that degradation must accompany the elimination of this hydrogen ion; it is not therefore possible to regard them as normal amphiprotic solvents. The position is rather similar for certain solvents which are classified with the amphiprotic solvents, such as acetonitrile CH_3CN, the protogenic character of which has been well shown, but of which the conjugate base CH_2CN^- is not stable and as soon as it is formed undergoes a consecutive transformation by its action on the CH_3CN molecules. For this reason, the distinction between the two groups of solvent is to some extent artificial in many cases.

3. Solvents neither protogenic nor protophilic in character are called 'aprotonic solvents'. Typical examples are phosphorus oxychloride and liquid sulphur dioxide. Some of these are practically inert: benzene, chlorobenzene, chloroform, carbon tetrachloride, etc.

Solvents which have been classified within the preceding category, but the (eventual) protogenic and protophilic characters of which are very weak and do not operate in practice, are frequently classified amongst aprotonic solvents.

It will be noted that the principal solvents quoted here are molecular solvents. It is amongst these latter that protolysis phenomena are most frequently observed. But they may equally well appear in ionized media (molten salts). For our immediate purposes, we will consider only molecular solvents and will deal with protolysis in molten salts only in the last part of this chapter.

In addition, it must be noted that we can introduce into a solution either molecular solutes corresponding to the definition of acids or of bases, or ionized solutes (ionophores) of which one of the ions (eventually both) corresponds to the definition. In this latter case, we shall be dealing with ionic acids or bases; if the solvent dielectric constant is sufficiently high, dissociation of ion-pairs is practically complete, the acidic or basic ion is free from any association with the counter-ion and this latter plays no part in the protolysis phenomena; but if the dielectric constant is small, the acidic or basic ion remains, at least partially, associated in the state of an ion-pair with the counter-ion, which changes the protolysis phenomena in a manner which we shall consider in Section II.3. Before that, we will consider only the case of solvents of high dielectric constant, where no ion-pair formation is to be allowed for (up to ε=about 20 for very dilute solutions).

We will take a special interest in the case of amphiprotic solvents, similar to water in that respect, and where for that reason the treatment approaches most closely to classical Brønsted treatment for aqueous solutions.

II.1. Acidity in a Molecular Solvent of High Dielectric Constant

AUTOPROTOLYSIS EQUILIBRIUM IN AN AMPHIPROTIC SOLVENT

We will use the symbol HS for the protogenic molecule (monomer) of such a solvent, thus showing the ionizable hydrogen atom which this molecule can eventually give up.

Due to the amphoteric nature of the HS molecules, there is set up, especially in the pure state, an equilibrium of exchange of H^+ between molecules acting as acid and others acting as base; we symbolize this equilibrium, known as autoprotolysis or self-ionization of the solvent, by:

$$2HS \rightleftharpoons H_2S^+ + S^- \quad \text{(II-5)}$$

H_2S^+ represents the solvated proton (by HS) and S^- the anionic conjugate base of the solvent HS.

This equilibrium is characterized by the autoprotolysis constant K_S which connects the activities of the three constituents of the equilibrium. Indeed, when the solvent is only slightly dissociated in this reaction and if it is practically pure, we can assume that the activity of the HS molecules is very nearly that which they would possess if they were alone, and allow it the value unity. In this way, the autoprotolysis constant is given by:

$$K_S = (a_{H_2S^+})\,(a_{S^-})\,. \quad \text{(II-6)}$$

Hence the name 'solvent ionic product', which is also given to it.

Under the conditions where we can assume that the solution is ideal and regard activity as equal to concentration, this becomes:

$$K_S = [H_2S^+]\,[S^-]. \tag{II-7}$$

The values of $pK_S = -\log K_S$, for some amphiprotic solvents, are collected in Table IV. For a given solvent, pK_S varies with temperature; for example, water is more dissociated at high temperatures than at low (pK_S falls with T).

In a rigorously pure amphiprotic solvent, the rule of electroneutrality requires that:

$$[H_2S^+] = [S^-] = K_S^{1/2}. \tag{II-8}$$

The concentration of free ions (which determines the ionic strength, since $I_c = K_S^{1/2}$) becomes smaller as pK_S increases.

Autoprotolysis constant of a mixture of two solvents

It is also of interest to consider, in addition to pure amphiprotic solvents, mixtures of solvents which behave similarly to an amphiprotic medium.

The hydrogen ion transfer, from which results the self-ionization of the medium, can here take place according to two different mechanisms:

1. The transfer takes place between the amphoteric molecules of one only of the two constituents, as if it were alone, whilst the other acts as an inactive diluent, and participating only, subsequently, in the solvation of the ions resulting from the self-ionization. The value of pK_S* is nevertheless variable with the relative proportions of the constituents of the mixture; in fact, when the proportion of inactive constituent increases, in principle the activity of the active constituent decreases (it becomes less than unity) – causing, all other things being equal, an increase in pK_S – , those of the ionic substances in solution are also modified as a result of the modifications of solvation.

Thus, in mixtures of water and an inactive miscible organic solvent, such as acetone or 1,4-dioxane, the autoprotolysis constant pK_S of water increases considerably with the proportion of organic solvent (experimental values of pK_S are given in Table IV).

2. The transfer can also take place between acidic molecules of one of the constituents and basic molecules of the other (the two solvents mixed are active, but not necessarily amphiprotic). The ionization of the mixtures is then

* K_S is still defined as in the case of a pure solvent – the product of the activities of the solvated proton and the conjugate anion base of the molecule of the constituent which gives up its proton – the activity of the solvent which intervenes in the self-ionization equilibrium, determined by the composition of the mixture, being included in its value.

TABLE IV

Value of the autoprotolysis constant (ionic product) K_s for various amphiprotic solvents (at 25 °C unless stated to the contrary).

Solvent HS	Cation H_2S^+	Anion S^-	$pK_s = -\log K_s/mol^2 l^{-2}$
Acetamide	$CH_3CONH_3^+$	CH_3CONH^-	14.6 (at 100 °C) [1]
Acetonitrile	CH_3CNH^+	CH_2CN^-	19.5 [2]
Acetic acid	$CH_3CO_2H_2^+$	$CH_3CO_2^-$	14.5 [3]; 15.2 [4]
Hydrofluoric acid	H_2F^+	F^-	10.7 (at 0 °C) [5]
Formic acid	$HCO_2H_2^+$	HCO_2^-	6.2 [6 7]
Hydrosulphuric acid	H_3S^+	HS^-	32.6 (at —78 °C) [8]
Sulphuric acid	$H_3SO_4^+$	HSO_4^-	2.9 [9]
Ammonia	H_4N^+	H_2N^-	27.7 [10]; 32 (at—60 °C) [11]
Dimethylsulphoxide	$C_2H_6SOH^+$	$C_2H_5SO^-$	33.3 [12]
Water	H_3O^+	HO^-	14.0 [13]; 15.0 (at 0 °C)[14]; 12. 3(at 100 °C) [15]
Ethanol	$C_2H_5OH_2^+$	$C_2H_5O^-$	19.1[16]; 18.9 [17]
Ethanolamine	$^+H_3NC_2H_4OH$	$H_2NC_2H_4O^-$	5.2 [18]; 5.7 [19]
Formamide	$HCONH_3^+$	$HCONH^-$	16.8 [20]
Hydrazine	$N_2H_5^+$	$N_2H_3^-$	~13 [21]
Methanol	$CH_3OH_2^+$	CH_3O^-	16.7 [22]

1-Propanol		$C_3H_7OH_2^+$	$C_3H_7O^-$	19.4 [23]
2-Propanol		—	—	20.8 [24]
Water-acetone mixtures		H_3O^+	HO^-	[24]
weight % acetone:	10			14.3
	25			15.0
	44			15.7
	70			16.5
	90			19.7
Water-acetic acid mixtures		H_3O^+	$CH_3CO_2^-$	[25]
weight % acetic acid:	5			4.85
	11			4.5
	22			4.3
	42.7			4.25
	62.4			4.45
	81.3			5.5
	90.6			6.75
	95			8.6

According to: [1] Guiot and Trémillon, 1968; [2] Romberg and Cruse, 1959; [3] Bruckenstein and Kolthoff, 1956; [4] Durand, 1970; [5] Kilpatrick *et al.*, 1956; [6] Hammett and Deyrup, 1932; [7] Mukherjee, 1957; [8] Jander and Schmidt, 1943; [9] Gillespie, 1950; [10] Coulter *et al.*, 1959; [11] Herlem, 1967; [12] Courtot-Coupez and Le Demezet, 1969; [13] Harned, 1925; [14] Harned *et al.*, 1933; [15] Perkovets and Kryukov, 1969; [16] Hartley *et al.*, 1932; [17] Schaal *et al.*, 1960; [18] Schaal and Masure, 1956; [19] Schaap *et al.* 1961; [20] Verhoek, 1936; [21] Vieland and Seward, 1955; [22] Hartley *et al.*, 1929; [23] Schaal and Teze, 1961; [24] Braude and Stern, 1948; [25] Le Ber, 1970.

TABLE IV (continued)

Solvent HS		Cation H_2S^+	Anion S^-	$pK_s = -\log K_s/\text{mol}^2\text{l}^{-2}$
Water-DMSO mixtures		H_3O^+	HO^-	[26]
weight % DMSO:	10.8			14.3
	21.3			14.55
	31.7			14.9
	51.2			15.9
	70.5			17.8
	80.4			19.3
	91			22.05
	95.5			24.0
Water-1,4-dioxane mixtures		H_3O^+	HO^-	[27]
weight % dioxane:	20			14.6
	45			15.75
	70			17.85
Water-ethanol mixtures				[28]
weight % ethanol:	20			14.35

	35			14.6	
	50			14.9	
	65			15.3	
	80			15.9	
Water-ethanolamine mixtures		$^{+}H_3NC_2H_4OH$	HO^-		[29]
weight % ethanolamine:					
	10			4.0	
	30			3.5	
	50			3.7	
	70			4.4	
	90			5.0	
Water-hydrazine mixtures:		$H_5N_2^+$	HO^-		[30]
weight % H_4N_2:	5			5.4	
	10			5.0	
	25			4.6	
	50			5.4	
	65			5.9	
	85			8.1	
	93.5			11.0	

According to: [26] Halle, Gaboriaud and Schaal, 1969; [27] Harned and Fallon, 1939; [28] Gutbezahl and Grunwald, 1953; [29] Grall, 1966; [30] Bauer, 1965.

greater (pK_S smaller) than that of the pure constituents; a minimum of pK_S is produced for a determined composition.

Let us consider for example the case of mixtures of water and hydrazine. The basic strength of hydrazine is greater than that of water, but the acidic strength of water is greater than that of hydrazine. In consequence, the H^+ ion transfer takes place in mixtures of H_2O and H_4N_2 according to the equilibrium:

$$H_2O + H_4N_2 \rightleftharpoons H_5N_2^+ + HO^- . \qquad \text{(II-9)}$$

The values of pK_S* for mixtures in variable proportions are given in Table IV.

Similar effects are observed in mixtures of water and ethanolamine (more basic but slightly less acidic than water), water and pyridine (more basic than water but not acidic), water and dimethylformamide (slightly more basic than water but in practice non-acidic), water and acetic acid (more acidic but less basic than water), etc.

In mixtures of water and sulphuric acid (much more acidic and much less basic than water), the relative basic and acidic strengths of the two constituents are such that H^+ transfer between the H_2SO_4 and H_2O molecules, as given by the equilibrium:

* K_S is defined by the relationship $K_S = (a_{H_5N_2^+})(a_{HO^-})$. The value of K_S is dependent on the activities of the solvents H_2O and H_4N_2, these latter being practically constant in a mixture of given composition.

$$H_2SO_4 + H_2O \rightleftharpoons H_3O^+ + HSO_4^- \qquad \text{(II-10)}$$

is displaced in the direction of a practically quantitative self-ionization. We can no longer define a constant ionic product; the autoprotolysis constant must be governed by the activities of the four constituents of the equilibrium. Further, the ionic strength of these mixtures is very large, and for this reason they are similar to ionized media.

3. In certain mixtures of two solvents with almost equal acid-base properties, the H^+ transfer can call into play the molecules of the two constituents simultaneously, both to give up H^+ ions and to fix and solvate them.

This is particularly true for mixtures of water and an alcohol ROH (amongst the first terms of the series), for which we can put the simultaneous equilibria:

$$\left\{\begin{array}{ll} ROH + HO^- \rightleftharpoons H_2O + RO^- & \text{(II-11)} \\ ROH + H_3O^+ \rightleftharpoons H_2O + ROH_2^+ & \text{(II-12)} \\ \left\{\begin{array}{l} 2H_2O \rightleftharpoons H_3O^+ + HO^- \\ \text{or} \\ 2ROH \rightleftharpoons ROH_2^+ + RO^- . \end{array}\right. & \begin{array}{l} \text{(II-13)} \\ \\ \text{(II-14)} \end{array} \end{array}\right.$$

This means that autoprotolysis gives rise simultaneously to anions HO^- and RO^- (conjugate bases of the mixture constituents), in a proportion determined by that of the two solvents – according to the mass action law applied to the equilibrium (II-11):

$$\frac{[RO^-]}{[HO^-]} = K \frac{a_{ROH}}{a_{H_2O}}. \qquad \text{(II-15)}$$

In the same way, protons have also appeared combined with H_2O and ROH, also in a proportion fixed by that of the mixture:

$$\frac{[ROH_2^+]}{[H_3O^+]} = K' \frac{a_{ROH}}{a_{H_2O}}. \tag{II-16}$$

In consequence, we can here define a constant 'apparent' ionic product, expressed as follows:

$$K_S = ([H_3O^+] + [ROH_2^+])([HO^-] + [RO^-]). \tag{II-17}$$

The activities of H_2O and ROH have an influence on the value of K_S.

In the pure mixture, the total concentration of solvated protons is equal to the total concentrations of the anions HO^- and RO^-.

Note. – In a general way, we can formulate the case of a mixture of two amphiprotic molecular solvents HS and HS', in which self-ionization remains small (in such a manner that the activities of the molecular constituents of the mixture can be taken as constant), by an expression for an apparent ionic product defined as in the special case of water-alcohol mixtures:

$$K_S = ([H_2S^+] + [H_2S'^+])([S^-] + [S'^-]) \tag{II-18}$$

with, in the pure mixture:

$$[H_2S^+] + [H_2S'^+] = [S^-] + [S'^-] = K_S^{1/2}. \tag{II-19}$$

If the acidic and/or basic strengths of the two constituents differ greatly, one term of one and/or of the other of the two concentration sums is negligible in comparison with the term added to it, and we can again be brought to one or other of the two extreme cases previously considered in 1 or 2. For example, if HS' is both less acidic and considerably less basic than HS, we still have $[H_2S'^+] \ll [H_2S^+]$ and $S'^-] \ll [S^-]$, such that the apparent ionic product K_S is brought to that of the single constituent HS. If HS' is much less acidic but much more basic than HS, we have on this occasion $[H_2S^+] \ll$ $\ll [H_2S'^+]$ and $[S'^-] \ll [S^-]$, and the ionic product takes the form $K_S = [H_2S'^+]\ [S^-]$.

Solvolysis of dissolved acids and bases

When an acidic or basic solute is introduced into a solution, it is first subjected, by the solvent molecules around it, to the effects of solvation. But when they themselves are endowed with acid-base properties, the solvent molecules are able to exert a further action of solvolysis on the solute (an action of the solvent molecules tending to dissociate the solute and to transform it chemically), corresponding to protolysis phenomena.

In all amphiprotic solvents, these phenomena are similar to hydrolysis of acids and bases in water. Thus, protolysis of an acid (symbolized by HA), in accordance with an equilibrium of type:

$$HA + HS \rightleftharpoons A + H_2S^+ \qquad \text{(II-20)}$$

appears as a dissociation of this latter, in the conjugate base A and solvated proton (symbolized here by H_2S^+, but we might more simply write H^+), which depends both on the nature of the acid under consideration and on that of the solvent. The acids which are practically completely dissociated are called 'strong' acids; 'weak' acids are only partially (and often very slightly) dissociated and are characterized by their dissociation constant or acidity constant K_A (or $pK_A = -\log K_A$) obtained from the following expression – having regard to the fact that in a sufficiently dilute solution the activity of the solvent HS remains practically invariable and can be taken as equal to unity:

$$K_A = \frac{(a_A)\,(a_{H_2S^+})}{a_{HA}}. \tag{II-21}$$

An acid is weaker as its dissociation is less, hence when the value of pK_A is greater.

Protolysis of bases (symbolized by A) in accordance with an equilibrium of type:

$$A + HS \rightleftharpoons HA + S^- \tag{II-22}$$

is shown by the appearance of the anion S^-, the conjugate base of the solvent. When protolysis is complete, the base is known as a 'strong base'. Weak bases are characterized by the basicity constant K_B (or $pK_B = -\log K_B$):

$$K_B = \frac{(a_{HA})\,(a_{S^-})}{a_A} \tag{II-23}$$

the value of which depends both on the base in question and on the nature of the solvent. A base is weaker as the value of pK_B is larger.

It is easily seen that the value of K_B is connected with that of K_A for the conjugate acid, by the relation:

$$K_A K_B = K_S, \quad \text{or} \quad \mathrm{p}K_A + \mathrm{p}K_B = \mathrm{p}K_S. \qquad \text{(II-24)}$$

From this, the value of pK_A is sufficient to characterize the acid-base couple in the solvent under consideration. The acid is stronger and the conjugate base weaker as the value of pK_A is smaller.

To sum up, any acid in solution is subjected to a dissociation, variable according to its nature and that of the solvent, if the latter has a protophilic character (basic), that is it 'solvates' the proton (the most frequent case). In the same way, any base in solution is subjected to a protolysis, variable according to its nature and that of the solvent, provided that this latter has a protogenic character (acidic; protic solvents only).

Definition of pH and limitation on the pH scale in an amphiprotic solvent

Two important features result from the fact that the solvent is both protophilic and protogenic and from the equilibria established with it.

1. The ionic particle brought into play in the acid-base phenomena, the hydrogen ion H^+, exists in all solutions

in the state of being solvated by the solvent molecules; we can therefore define the concentration and the activity of this characteristic species.

2. This concentration (or this activity) is variable according to the nature and concentration of the acids and/or bases introduced into the solution. In fact, due to the dissociation equilibrium established for any dissolved acid, the introduction of an acid into the solution must necessarily increase the concentration of solvated H^+ ions (with regard to their initial concentration $K_S^{1/2}$ in pure solvent); an acid is thus a compound which brings solvated H^+ ions into solution. On the other hand, due to the protolysis equilibrium of bases in a protogenic solvent, the introduction of a base into the solution must necessarily increase the concentration of S^- anions (a base is thus a body which brings anions S^- of the solvent into solution, in a greater quantity than those already present in the pure solvent; this coincides with the definition of a base according to Arrhenius for aqueous solutions: donor of hydroxide ions HO^-). But since the concentrations of S^- and of H_2S^+ remain connected by the constant ionic product, the increase in $[S^-]$ must decrease $[H_2S^+]$ and the introduction of a base causes a reduction in the concentration of solvated H^+ ions in comparison with their initial concentration in the solvent.

It is consequently logical to use, as a measure of the state of the solution as far as acidity is concerned, a quantity connected with this concentration of solvated H^+ ions.

For aqueous solutions, Sørensen (1909) first suggested for consideration the value of the negative logarithm of the hydrated H^+ ion concentration; this value is known as pH.* As a means of experimental measurement, it was first suggested that the electromotive force of a cell comprising a reference electrode and a hydrogen electrode (Pt; H_2 gas under fixed pressure, in general 1 atm, and solution saturated of hydrogen under these conditions) should be used.

This electromotive force E, after elimination of the liquid-junction potential, is a function of the value of the only variable activity in the cell, that of the H^+ ions:

$$E = \text{constant} - \frac{2.303RT}{F} \log a_{\mathrm{H}^+} . \tag{II-25}$$

When it was realized that, in this type of measurement, the activities and not the concentrations are relevant, a modification to the definition of pH was suggested so as to remain in agreement with the suggested method of measurement (Sørensen and Linderstrøm-Lang, 1924):

$$\mathrm{pH} = -\log a_{\mathrm{H}^+} = -\log(\gamma_{\mathrm{H}^+} [\mathrm{H}^+]) = \frac{F}{2.3RT}(E - E^0) \tag{II-26}$$

* Sørensen first wrote this as p_{H}, but the notation pH was gradually adopted, and has now become the one in most common use today.

(E° is the value of the electromotive force at pH=0, and γ_{H^+} the activity coefficient of the hydrated proton). Potential measurements of a glass electrode, which are those most commonly carried out today, also agree with relation (II-26). This definition is therefore the one generally adopted at the present day, in preference to the original definition.

There is only a small difference between the values corresponding to the two definitions when the ionic strength of the solution is small (γ_{H^+} nearly equal to unity).

The same definition may be retained for the acidity measurements in all other solvents:

$$\mathrm{pH}(\mathrm{H}S) = -\log a_{\mathrm{H}_2S^+} = -\log(\gamma_{\mathrm{H}_2S^+}[\mathrm{H}_2S^+]) \tag{II-27}$$

where pH(HS) is the pH in the solvent HS under consideration.

pH forms a logarithmic scale of activity of the solvated hydrogen ions. Low values of pH correspond to the strongest activities, and high values to the weakest activities.

Some characteristic features of the 'pH scale' in amphiprotic solvents are as follow:

1. In the pure solvent, we have according to relation (II-8):

$$\mathrm{pH} = \tfrac{1}{2}\mathrm{p}K_S. \tag{II-28}$$

This value defines 'neutrality' for solutions in the solvent HS under consideration.

2. Since acids are substances which lead to an increase in the H_2S^+ ion concentration, we must have $pH < \frac{1}{2}pK_S$ for any acidic solution. This range of pH values defines acidic media in the solvent under consideration.

The pH value becomes less as the acid introduced is more concentrated and more dissociated, that is stronger (smaller value of pK_A). Strong acids lead to the most acidic solutions, since they are completely dissociated.

We can adopt approximately as the limit of 'dilute solutions' concentrations in the neighbourhood of 1 mol l^{-1} or 1 mol kg^{-1}. Under these conditions, the minimum pH value for a solution is near to zero (or slightly negative values).

3. Since bases are substances which lead to a fall in the H_2S^+ ion concentration, we must have for any basic solution $pH > \frac{1}{2}pK_S$. This range of pH values defines basic media in the solvent under consideration.

The pH value is higher as the base introduced is more concentrated and stronger (smaller value of pK_B). Strong bases, in particular the fully ionized salts of S^-, lead to the most basic solutions. The maximum value of pH for a solution is close to $pH = pK_S$ (or slightly more for base concentrations greater than 1 mol l^{-1} or 1 mol kg^{-1}).

Thus we may regard the extent of the pH scale in the solvent concerned at a number of units close to pK_S, the value $pH=\frac{1}{2}pK_S$ (neutrality) corresponding to the middle of the acido-basic range. The possible pH variation in a solvent is therefore greater as pK_S is greater, that is as the solvent itself is less ionized.

The diagram in Figure 1 summarizes the characteristics of the pH scale.

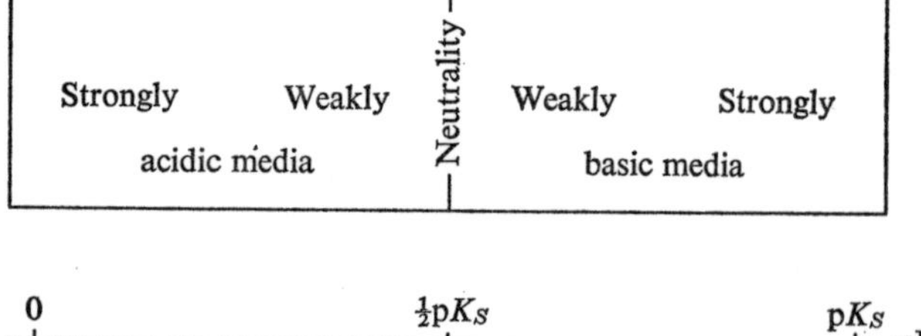

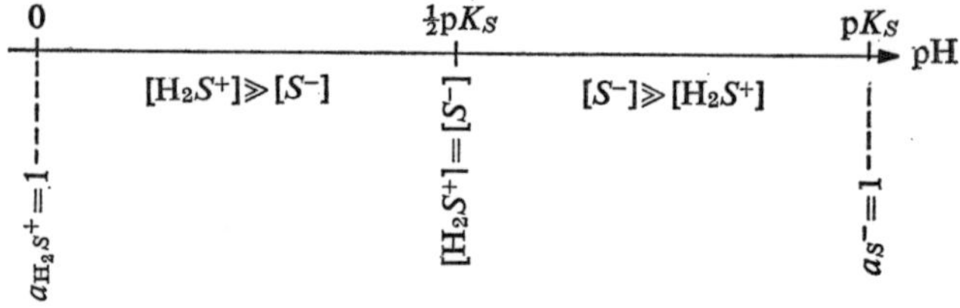

Fig. 1. pH scale, with acidic and basic ranges, in an amphiprotic solvent (HS).

Examples

1. Let us first recall that in an aqueous solution the extent of the pH scale is about 14 units at ordinary temperatures (the concentration of hydrated H^+ can vary from about 10^{-14} to more than 1 M). Neutrality corre-

sponds to pH=7, acidic media to pH<7 and basic media to pH>7. The usual strong acids are perchloric, hydrochloric, nitric, and sulphuric. The strong bases are the alkali hydroxides; alcoholates and amides of alkali cations are completely hydrolyzed and therefore they also behave as strong bases.

2. In an alcohol ROH – particularly methanol or ethanol – , the autoprotolysis equilibrium being symbolized by:

$$2ROH \rightleftharpoons ROH_2^+ + RO^- \qquad \text{(II-29)}$$

ROH_2^+ represents the solvated proton and the pH is defined as the negative logarithm of the activity of this species.

The extent of the pH range is about 19 units for ethanol (neutrality at pH≈9.5) and nearly 17 units for methanol (neutrality at pH≈8.3).

If perchloric acid is still a strong acid in methanol or ethanol, hydrochloric acid becomes weak in these two solvents; $pK_A=1.2$ in MeOH and 2.1 in EtOH; acetic acid is appreciably weaker (less dissociated) than in aqueous solution: $pK_A=9.7$ in MeOH, and 10.4 in EtOH.

In these circumstances, the strong bases are alkali alcoholates (anion RO^-, conjugate base of ROH). Ammonia, pyridine, and aniline are even weaker bases than in aqueous solution; the pK_B values for these bases are respectively 8.7, 14.4 and 13.4 in ethanol, compared with 4.8, 8.8 and 9.4 in water. H_2O also shows a very slightly basic character in ethanol ($pK_A=0.3$).

3. In liquid ammonia, since the autoprotolysis equilibrium is symbolized by:

$$2NH_3 \rightleftharpoons NH_4^+ + NH_2^- \qquad \text{(II-30)}$$

the NH_4^+ ion represents the proton solvated by this solvent; thus we can define:

$$pH = -\log a_{NH_4^+} \,. \qquad \text{(II-31)}$$

At —60 °C, the extent of the pH scale is about 32 units; it is therefore much wider than in water. Neutrality corresponds to $pH \approx 16$, acidic media to $pH < 16$ and basic media to $pH > 16$.

Many acids which are weak in aqueous solution become strong in ammonia, notably acetic acid. Extremely weak acids, showing no acidic properties in aqueous solution, are greatly dissociated in ammonia.

The conjugate base of NH_3 is the amide anion NH_2^-; potassium amide is therefore the strong base in liquid ammonia. Alcoholates and hydroxides are weak bases. The pH of a solution of 10^{-2}M amide is close to 30, whilst that of a solution of 10^{-2}M hydroxide is only about 24.5.

4. In molten acetamide, the autoprotolysis equilibrium:

$$2CH_3CONH_2 \rightleftharpoons CH_3CONH_3^+ + CH_3CONH^- \qquad \text{(II-32)}$$

enables us to define the solvated proton $CH_3CONH_3^+$, and the strong base which is here the acetamide anion CH_3CONH^-.

At 100 °C, the extent of the pH scale is about 14.6 units; neutrality corresponds to pH=7.3. Hydrochloric and nitric acids are strong acids in molten acetamide; acetic acid is even weaker than in water: $pK_A=5.8$ in place of 4.95 in aqueous solution in the neighbourhood of 100 °C.

5. In hydrofluoric acid, the autoprotolysis equilibrium is symbolized by:

$$2HF\,(\text{or } H_2F_2) \rightleftharpoons H_2F^+ + F^- . \qquad \text{(II-33)}$$

The fluoride anion F^- is the strong base, and all bases such as NH_3, $CH_3CO_2^-$, HO^- and pyridine, which liberate F^- quantitatively by reaction with HF, also behave like strong bases in hydrofluoric acid. Very weak bases, showing no basic character in aqueous solution, can undergo protonation in HF: Cl^-, alcohols, ethers are bases strong in practice; water ($pK_B=1.1$), carboxylic acids, naphthalene ($pK_B=4.0$), even benzene ($pK_B=9.4$) are weak bases.

On the other hand, acids which are weak in aqueous solution no longer exhibit their acidic character here, and strong acids become weak ($HClO_4$, H_2SO_4) and even sometimes extremely weak (HCl).

The extent of the pH scale at 0°C is about 11 units; neutrality corresponds to pH$\approx$5.8. It is very difficult to obtain the most acidic media, due to the lack of acids which are strong enough in such a solvent.

6. Pure sulphuric acid presents a close analogy with hydrofluoric acid. In this solvent, the autoprotolysis equilibrium is:

$$2H_2SO_4 \rightleftharpoons H_3SO_4^+ + HSO_4^- \qquad \text{(II-34)}$$

and the strong base is therefore the hydrogen sulphate ion HSO_4^-. All bases, weak in aqueous solution, which liberate HSO_4^- quantitatively by reaction with H_2SO_4, behave here as strong bases, and bases which exhibit no basic properties in water do so in sulphuric acid: the anions Cl^-, NO_3^-, ClO_4^-, molecules of carboxylic acids, of alcohols, ethers, etc; even water itself is in practice a strong base under these circumstances.

On the other hand, as in hydrofluoric acid, very few acids can exhibit their acidic character: HCl is no longer acid, and $HClO_4$ is only a very weak acid; one of the strongest acids is disulphuric acid $H_2S_2O_7$ (obtained by dissolving SO_3 in H_2SO_4), which is characterized by $pK_A=1.5$. All these acids are strong in aqueous solution.

The extent of the pH scale is very small, since sulphuric acid when pure is already very much ionized: it is only about 3 units, between $[H_3SO_4^+]=1$ M and $[HSO_4^-]=1$ M.

The situation in sulphuric acid is more complicated than

in the preceding solvents because of the intervention of another auto-dissociation of the H_2SO_4 molecules, corresponding to dehydration of sulphuric acid into disulphuric acid; having regard to the strong basic character of water and the strong acidic character in practice of dilute disulphuric acid, we must write this equilibrium:

$$2H_2SO_4 \rightleftharpoons H_3O^+ + HS_2O_7^- \qquad \text{(II-35)}$$

with:

$$K'_S = [H_3O^+][HS_2O_7^-] = 10^{-3.1}\ \text{mol}^2\ \text{l}^{-2}. \qquad \text{(II-36)}$$

Mixtures of two solvents

In mixtures of solvents such as those previously defined, and in particular in hydro-organic mixtures, the pH is defined as the negative logarithm of the *total* activity of the solvated proton, whether the H^+ ion is combined with one only of the constituents or with both simultaneously.

The extent of the accessible pH scale is limited to the values reached for the greatest concentrations of the base conjugate to one of the constituents – the stronger acid of the two – or of the two conjugate bases together when these latter exist together in nearly the same proportions – that is when the two solvents which form the mixture have almost the same acidic strengths. From the relation (II-18), defining the apparent ionic product of such a medium, we have pH=pK_S when $[S^-] + [S'^-] = 1$ M (or 1 mol kg^{-1}).

From (II-19), neutrality still corresponds to $pH = \frac{1}{2}pK_S$.

Let us recall that the ratio $[S'^-]/[S^-]$ (or the proportion $[S'^-]/\{[S^-]+[S'^-]\}$) is determined by the relative acidic strengths of the two constituents HS and HS' and by their proportions in the mixture. It therefore remains unchanged whatever the value of pH, that is whatever the acids and bases dissolved and their equilibrium concentrations. The same applies to the ratio $[H_2S'^+]/[H_2S^+]$.

The extent of the accessible pH range varies when the proportions of the two solvents are varied, depending on the modifications to the autoprotolysis constant pK_S value. Let us consider for example hydro-organic mixtures; compared with purely aqueous solutions, the extent of the pH scale is increased by addition of an inactive solvent (acetone, 1,4-dioxane); it falls and passes through a minimum when the organic constituent is active, either more acidic or more basic than water.

In addition, the strengths of acids and bases, that is the values of pK_A of the acid-base couples, are also modified, progressing from the value in pure water to that in the other pure constituent, when the composition of the mixture is progressively varied. Acids (or bases) which are strong in water and weak in the other constituent, or vice versa, appear weak within a certain range of composition and strong in the remainder of the range.

With regard to mixtures of water and sulphuric acid (and analogous mixtures such as water $+HNO_3$, water$+$ $+HCl$, etc), we note that in accordance with what has previously been said about their state of ionization, the pH cannot be varied in practice (without a considerable modification to the composition of the mixture), that is the pH scale is reduced to one point. In fact, the activity of solvated protons (H_3O^+) is large, imposed by the composition of the mixture, and hence impossible to modify to any great extent by addition of dilute acids or bases. In consequence, acidic or basic solutes in such a mixture are in a state that cannot be modified from the acidity aspect; no acid-base transformation (other than those taking place on dissolving by solvolysis) can take place there.

Acid-base reactions and pH variation during these reactions

Having regard to solvolysis, acid-base couples are classified in three categories in amphiprotic solvents.

(a) Strong acids, practically or completely dissociated, all give rise quantitatively to the H_2S^+ species, which is in their case the only reactive acid group. Their conjugate bases, unable to combine with the H^+ ion because the acidic forms do not exist, have no basic properties in the solvent under consideration ('nul' bases).

(b) On the other hand, all strong bases give rise quantitatively to the S^- species, the only reactive group in their case. Their conjugate acid forms, practically incapable of dissociation, have no acidic properties in the solvent under consideration ('nul' acids).

TABLE V

Values of the constant pK_A for some acid-base couples in various solvents (at 25 °C unless stated to the contrary)

Solvent	Acid–base couple										
	Hydrochloric acid–chloride	Picric acid–picrate	Benzoic acid–benzoate	Acetic acid–acetate	Phenol–phenate	Ammonium–ammonia	Anilinium–aniline	Piperidinium–piperidine	Pyridinium–pyridine	H_3O^+–water	Water–HO^-
Water	SA	0.8	4.2	4.75	9.9	9.25	4.6	11.1	5.2	–	–
Water (at 98 °C)	SA		4.45	4.95		7.45		9.15		–	–
Acetamide (at 98 °C) [1]	SA		5.55	5.8	10.35	4.9		7.1	1.45		SB
Acetonitrile [2, 3]	8.1	8.9	12.0							2.7	
Formic acid [4]	1.4				SB	SB	SB	SB	SB	1.4	SB
Ammonia (at —60 °C) [5]	SA	SA	SA	SA	3.5	–	SA		SA	SA	18.9
Dimethylformamide [6]	3.4	1.2	10.8	11.1	18	9.5		10.1	3.0		
Dimethylsulphoxide [7]			10.9	12.6	16.9		3.7	10.6	3.5		28.2
Ethanol [8]	2.1	4.1	10.1	10.4		10.4	5.7		4.7	0.3	
Formamide [9]	SA	1.2	6.2	6.8			4.1		4.5		
HMPA [10]	SA		8.25	11.9	14.15	10.3	1.95	8.3	1.0	SA	
Hydrazine [11]	SA	SA	SA	SA	3.2					SA	
Methanol [8]	1.2	3.8	9.4	9.7	14.0	10.8	6.0				
Nitrobenzene [12]		7.4					4.8		8.0		
Water-1,4-dioxane mixtures [13]											
dioxane weight %: 20				5.3							

	80		9.75				
Water-ethanol mixtures [14,15,16]:							
ethanol weight %:	20	4.75	5.15	9.0	4.4	4.9	
	35	5.31	5.4	8.8	4.15	4.45	
	50	5.8	5.85	8.6	3.9		4.2
	65		6.2	8.45	3.8		
	80		6.6		3.75		
	95		7.6		4.2		
Acetone-water mixtures [15]:							
acetone weight %	18		5.2		4.4		
	36		5.85		4.1		
	46		6.25		3.9		
	56		6.75		3.75		
	66		7.4		3.6		
	76		8.3		3.45		
	88		9.7		3.6		
	94		11.5		3.7		
Water-DMF mixtures [16]:							
DMF volume %	20	4.6	5.05		4.25	4.7	
	30	4.8	5.25		4.1	4.45	
	40		5.45		3.9	4.2	
	50		5.8		3.7	3.9	
	60		6.25		3.5	3.6	
	70		6.85		3.2	3.3	
	80		7.8		3.0	3.05	

S.A = acid is strong; S.B. = base is strong.

According to: [1] Guiot and Trémillon, 1968; [2] Romberg and Cruse, 1959; [3] Kolthoff *et al.*, 1961; [4] Mukherjee, 1957; [5] Badoz-Lambling *et al.*, 1969; [6] Bréant and Demange-Guérin, 1969; [7] Courtot-Coupez and Le Demezet, 1969; [8] Kilpatrick *et al.*, 1940, 1953; [9] Verhoek, 1936; [10] Madic and Trémillon, 1968; [11] Vieland and Seward, 1955; [12] Kolthoff *et al.*, 1953; [13] Bell and Robinson, 1961; [14] Gutbezahl and Grunwald, 1953; [15] Douheret, 1967; [16] Reynaud, 1967.

(c) Couples consisting of a weak acid and a weak base, where the constituents are only slightly, or even not at all, solvolized and which exist in the solution in the solvated state, have a strength characterized by the constant pK_A.

We can thus note that the cation H_2S^+ is the strongest acid which can exist in the solvent H*S*, since stronger acids are completely solvolized, whilst the anion S^- is the strongest base which can exist in H*S*, since stronger bases also undergo complete protolysis on the part of the solvent.

From this, we can consider the following different types of acid-base reaction:

1. The reaction between a strong acid and a strong base leads to neutralization:

$$H_2S^+ + S^- \rightarrow 2HS. \qquad \text{(II-37)}$$

Characterized by the equilibrium constant pK_S, this is the most 'quantitative', that is the most complete acid-base reaction which can take place in the solvent under consideration. It causes the greatest variations in pH (between the initial solution of strong acid and the final solution of strong base, or vice versa), corresponding to the value of pK_S.

2. The reaction of a weak acid HA with a strong base is represented by:

$$HA + S^- \rightarrow A + HS. \qquad \text{(II-38)}$$

The transformation of HA into A becomes less quantitative and the variation in pH (between the initial solution of weak acid HA and the final solution of weak base A with an excess of strong base S^-) becomes smaller as the equilibrium constant pK_A increases, therefore when the acid is weaker.

3. In the same way, the reaction between a weak base and a strong acid, represented by:

$$A + H_2S^+ \rightarrow HA + HS \qquad \text{(II-39)}$$

becomes less quantitative as pK_A becomes smaller, that is as the base A is weaker.

4. The reaction between a weak acid HA and a weak base A′:

$$HA + A' \rightarrow A + HA' \qquad \text{(II-40)}$$

only takes place to an appreciable extent if the value of pK_A of the couple HA/A is less than that of the couple HA′/A′. The difference in pK_A determines the magnitude of the pH variation during the reaction.

Since the values of the acidity and basicity constants of the same acid-base couple differ with the nature of the solvent, it is self-evident that the reaction between a particular acid and base follows a different path in different solvents, and does not lead to the same variation in pH. By way of example, Figure 2 shows the curves of pH variation during the reaction between hydrochloric acid and ammonia in three different solvents. In water, HCl is a strong acid and NH_3 a weak base ($pK_A=9.2$); in formic acid, NH_3 is a strong base, whilst HCl becomes a weak acid ($pK_A=1.4$); finally, in dimethylformamide, HCl is a weak acid ($pK_A=3.4$) and NH_3 is a weak base ($pK_A=9.5$).

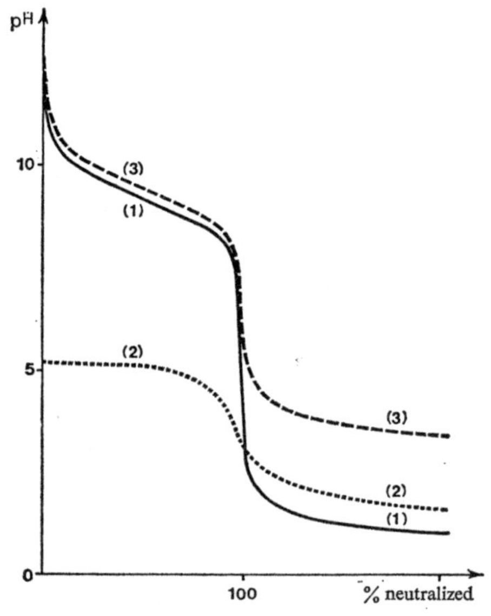

Fig. 2. Theoretical pH variation during neutralization by HCl of a 0.1 M ammonia solution in (1) water, (2) formic acid, and (3) dimethylformamide.

ACIDITY IN NON-PROTOGENIC MOLECULAR SOLVENTS

In comparison with amphiprotic solvents, the acid-base

phenomena in non-protogenic solvents exhibit the following differences:

1. The solvent does not give place to an autoprotolysis equilibrium.

2. If the solvent is protophilic (such as acetone, dioxane, tetrahydrofuran, DMF, HMPA, etc.), the dissolved acids are more or less dissociated; but the dissolved bases do not undergo protolysis. In consequence, there are strong acids but no strong bases (since there is no conjugate basic form of the solvent); each acid-base couple remains characterized by an acidity constant pK_A (but there is no basicity constant pK_B).

The pH scale, of which the origin still corresponds to unit activity of the solvated proton (HS^+, if S represents the protophilic solvent molecule), is in principle without limit towards the basic media. There is no point of neutrality.

HS^+ is the strongest acid that can exist, but there is no limitation to the strength of bases set by the solvent.

The pH of a solution containing no acid (that is to say the pure solvent and solutions of pure bases) remains in principle indeterminate.

3. If the solvent is not protophilic, it does not solvate the proton, and the dissolved acids are not dissociated. We can then no longer define the acidity constant pK_A.

Nevertheless, acid-base reactions can take place even in such a solvent, all these being of course of the type represented by the equilibrium (II-2), characterized by the constant (II-3). But the solvent no longer intervenes except by solvation of the species participating in this equilibrium.

By taking an acid-base couple as a reference point, it is possible to classify with regard to each other all the acid-base couples in such a solvent, by means of the values of the logarithms of the equilibrium constants for exchange of H^+ between the acid or base used as reference and the base or acid of the couple under consideration. This classification is similar to that drawn up in a protophilic solvent in accordance with the values of pK_A, the differences in pK_A corresponding to the values of the exchange constants mentioned above.

II.2. Acidity of Acceptors of the Anion given up by an Amphiprotic Solvent

Species other than H^+ ions are capable of combining with basic anions coming from the autoprotolysis system of the solvent when this latter is amphiprotic; this is particularly true for metallic ions. Those which have the property of being 'acceptors' of S^- anions, with which they then form 'complexes', cause, when dissolved in solvent HS, a displacement of the autoprotolysis equilibrium in the direction of liberation of solvated hydrogen ions, which gives them an acidic character.

These substances are not intrinsically acids. They exhibit no acidic character in a non-protogenic solvent. For this reason, they are not considered at the same time as the other acid-base couples.

Acidity of metallic ions

Let HS be an amphiprotic solvent in which a metallic cation M^{n+} is dissolved (the solute is the solvated cation M^{n+},aHS). By adding to this solution the strong conjugate base of the solvent, S^-, we can observe the formation of various combinations, successive complexes MS, MS_2, etc. (without indicating their ionic charge; the combination MS_n is molecular, lower complexes are cationic, and the later higher complexes are anionic*). This property has the following results:

(a) The incompletely dissociated compounds MS_x (almost undissociated if they are very stable) no longer behave as strong bases like the free S^- ion, but as weak bases of strengths varying with their stability. Combinations intermediate between M^{n+} and the higher complex are amphoteric in nature: basic with regard to

* Certain of these combinations, and particularly MS_n, may have a very low solubility and may therefore be precipitated when formed.

For simplicity, we only consider 'mono-nuclear' combinations, of type MS_x, ignoring 'poly-nuclear' groups of type M_yS_x which can also be formed under certain conditions.

M^{n+}, but acidic with regard to the higher complex (that containing the larger number of S^- ions).

(b) The solution of the metallic cation is acidic – pH less than $\frac{1}{2}pK_S$ – since the concentration of free H^+ (H_2S^+) is increased with regard to that in the pure solvent, due to the establishment of acid-base equilibria:

$$\begin{cases} M^{n+} + 2HS \rightleftharpoons MS + H_2S^+ \\ MS + 2HS \rightleftharpoons MS_2 + H_2S^+ \\ \cdot \quad \cdot \quad \cdot \quad \cdot \quad \cdot \quad \cdot \quad \cdot \quad \cdot \quad \cdot \end{cases} \tag{II-41}$$

The formation of several successive complexes has the effect of giving the metallic cation a polyacid character. Several acidity constants pK_A, corresponding to the equilibria of (III-41), for example:

$$K_{A.1} = \frac{[MS]\,[H_2S^+]}{[M^{n+}]} \tag{II-42}$$

characterize the strength of this apparent acid, as in the case of classic acid-base couples. We can also formulate this acid as $M(HS)^{n+}$ to draw attention to its proton-donor characteristic (if M^{n+} is not an acid in the sense of the Brønsted–Lowry definition, the solvated cation is).

It is evident that the acidic character of the metallic ion is dependent on the nature of the solvent; the various metallic ions do not have the same strengths in the different amphiprotic solvents; this depends on the stability of the complexes formed with the S^- anions.

A practical consequence of these protolysis phenomena dealing with metallic ions is that these latter can exist in the free state, just solvated, only if the solution is sufficiently acidic. In fact, the most acidic form $M(HS)_a^{n+}$ can predominate only if $pH \ll pK_{A.1}$; above this value, the various basic combinations appear and then become predominant.

When an intermediate form – generally the uncharged compound MS_n – is sparingly soluble, its formation brings with it a substantial fall in the apparent solubility of the metallic ion. This formation, and in consequence the insolubility of the metallic ion, occurs within a certain range of pH, determined by the values of the equilibrium constants and the solubility products. This effect of pH on the solubility of metallic ions has important applications, as much for analytical purposes as for preparational purposes.

Examples

The acidity of metallic ions in aqueous solution, resulting from the formation of hydroxo-complexes, is a well-known phenomenon. Most metallic hydroxides have small solubilities in water; their formation brings with it a precipitation of the metallic ion. The most acidic cations (Al^{3+}, Fe^{3+}, Zr^{4+}, Hg^{2+} ...)* are precipitated

* In a general way, the acidity of many metallic cations in solution increases in inverse proportion to their size, and increases especially with their ionic charge.

starting from low pH values; certain amphoteric hydroxides are redissolved in a strongly basic medium by formation of higher anionic hydroxo-complexes: $Al(OH)_3 \rightarrow$ aluminate $Al(OH)_4^-$, $Zn(OH)_2 \rightarrow$ zincate $Zn(OH)_3^-$, etc. Finally, we note that in some cases oxides and not hydroxides are formed: HgO, Ag_2O for example.

In a similar manner, we can consider the acidity of the same cations in various other amphiprotic solvents.

In liquid ammonia, metallic cations are rendered acidic by their ability to accept the amide ion NH_2^-, the strong base of the solvent. Limiting ourselves for simplicity to formation of metallic amides $M(NH_2)_n$, most frequently sparingly soluble in liquid ammonia as hydroxides are in water, the acid-base equilibrium here is:

$$M^{n+} \text{(solvated by } NH_3) + 2n\, NH_3 \rightleftharpoons$$
$$\rightleftharpoons M(NH_2)_n\,(s) + n\, NH_4^+ . \qquad \text{(II-43)}$$

Certain amides of low solubility are amphoteric and are therefore also redissolved in a strongly basic medium (in presence of potassium amide) by formation of higher anionic complexes: $AgNH_2 \rightarrow Ag(NH_2)_2^-$, $Zn(NH_2)_2 \rightarrow$ $\rightarrow Zn(NH_2)_4^{2-}$, $Al(NH_2)_3 \rightarrow Al(NH_2)_4^-$, etc.

We also observe, in certain cases, in place of formation of amides, that of imides (combinations with HN^{2-}) or of nitrides (combinations with N^{3-}), which are the equivalent of oxides (combinations with O^{2-}) in water; for example, the following precipitation reactions have been observed by increasing pH: $Pb^{2+} \rightarrow PbHN$ (*s*),

$Hg^{2+} \rightarrow Hg_3N_2$ (*s*), $Bi^{3+} \rightarrow BiN$ (*s*) (these nitrides are explosive).

The correspondence between combinations of hydroxide ions in water and amide ions in liquid ammonia, on one hand, of oxide ions in water and imide or nitride ions in ammonia, on the other hand, was particularly stressed by Franklin (1935). In the same manner that the oxide ion O^{2-} is a strong base in water and hence could not exist in the free state ($O^{2-} + H_2O \rightarrow 2\ OH^-$), but can exist in presence of water in the form of metallic compounds with sufficient stability to cause dissociation of the OH^- ions, the NH^{2-} ions (conjugate base of NH_2^-) and N^{3-} (conjugate base of NH^{2-}) are strong bases in liquid ammonia and do not exist there in the free state,

TABLE VI

Combinations, the formation of which inside an amphiprotic molecular solvent renders the metallic ions acidic

Solvent	Combinations of metallic ions with:
Acetamide	Acetamide ion CH_3CONH^-
Acetic acid	Acetate $CH_3CO_2^-$
Hydrofluoric acid	Fluoride F^-
Formic acid	Formate HCO_2^-
Hydrogen sulphide	Hydrogen sulphide ion HS^- and sulphide S^{2-}
Sulphuric acid	Hydrogen sulphate HSO_4^- and sulphate SO_4^{2-}
Alcohols ROH	Alcoholates RO^-
Ammonia	Amide ion NH_2^-, imide ion NH^{2-} and nitride N^{3-}
Water	Hydroxide OH^- and oxide O^{2-}

but in the form of sufficiently stable metallic compounds.

To summarize the different causes of acidity of metallic cations in a few amphiprotic molecular solvents in common use, Table VI gives the list of corresponding combinations the formation of which causes this effect of acidity.

OTHER SOLUTES BEHAVING AS ACIDS IN AMPHIPROTIC SOLVENTS

Alongside metallic ions and their compounds, there are numerous other compounds capable of exhibiting an acidic nature when dissolved in an amphiprotic solvent (and only in these latter).

This is the case, in water, of all the 'acidic anhydrides' (non-metal oxides), certain of which combine quantitatively with water ($SO_3 \rightarrow H_2SO_4$, $P_2O_5 \rightarrow H_3PO_4$, $As_2O_3 \rightarrow As(OH)_3$, $B_2O_3 \rightarrow B(OH)_3$, $(CH_3CO)_2O \rightarrow 2CH_3CO_2H$, etc.), but where others are only weakly hydrated and are nevertheless acids:

$$CO_2 + 2H_2O \rightleftharpoons HCO_3^- + H_3O^+$$
$$SO_2 + 2H_2O \rightleftharpoons HSO_3^- + H_3O^+ .$$

When dissolved in a non-protogenic solvent, these anhydrides exhibit no acidic character; dissolved in other, protogenic, solvents, they produce on the other hand various acids, of different strengths according to the nature of the solvent. Thus, sulphur trioxide leads to the following acids: in water, $SO_3 + H_2O \rightarrow H_2SO_4$; in sulphuric acid, $SO_3 + H_2SO_4 \rightarrow H_2S_2O_7$; in hydrofluoric acid, $SO_3+HF \rightarrow HSO_3F$; in liquid ammonia, $SO_3+NH_3 \rightarrow HSO_3NH_2$ (sulphamic acid, diacide).

In the same manner, compounds such as PF_3, SbF_5, PF_5, behave as acids in liquid hydrofluoric acid: $BF_3 \rightarrow HBF_4$, $SbF_5 \rightarrow HSbF_6$ (this is the strongest acid known in hydrofluoric acid: $pK_A=1.8$). In solution in liquid ammonia, boron trifluoride also behaves as a weak acid (H_3NBF_3).

Solvolytic reactions in amphiprotic solvents

In quite a general manner, 'solvolysis' is the decomposition of a substance when it dissolves, by the action of the solvent. Solvolysis is very often limited to a simple (ionic) dissociation of the substance into several solvated species; this is the case that we have previously considered, with HA acids; this is also what may take place for complex compounds, the stability of which varies with the nature of the solvent.

But it can also happen that solvolysis leads to a real transformation of the dissolved substance, giving rise to other solutes, of natures varying with the nature of the solvent. These changes are known as 'solvolytic reactions'. They most frequently involve an exchange of cations and anions between the solvated substance and the solvent molecules. This implies that the solvent is itself a donor of ions; this is the case for amphiprotic solvents H*S*, donors simultaneously of cations H^+ and of anions S^-.

Various solvolysis reactions are therefore produced in

amphiprotic solvents, consisting of a replacement of anionic groups of the solute by the S^- anions of the solvent. In consequence, there appear simultaneously solvated hydrogen ions or other solutes of weakly acidic character, in such a manner that these solvolytic reactions are related to the protolysis phenomena.

We know, for example, the hydrolysis phenomena in aqueous solution during which hydroxide OH^- or oxide ions O^{2-} given up by the water take part, and which take the place of anionic groups in the solute molecule: hydrolysis of metallic complexes, transformed into hydroxides or into mixed complexes (ex: $BiCl_3 \rightarrow BiOCl$, $ZrCl_4 \rightarrow ZrO^{2+}$ and $ZrOCl^+$, $UF_6 \rightarrow UO_2^{2+}$), also hydrolysis of compounds of non-metals and of organic compounds:

$$BCl_3 + 6H_2O \rightarrow B(OH)_3 + 3(H_3O^+ + Cl^-)$$

$$SO_2Cl_2 + 4H_2O \rightarrow SO_2(OH)_2 \text{ (or } HSO_4^- + H^+) + 2(H_3O^+ + Cl^-)$$

$$RCl + 2H_2O \rightarrow ROH + H_3O^+ + Cl^-$$

$$2RHgCl + H_2O \rightarrow 2RH + HgO(s) + HgCl^+$$

Corresponding 'ammonolysis' reactions take place in liquid ammonia, using NH_2^- ions in place of OH^-:

$$ZrCl_4 + 2NH_3 \rightarrow Zr(NH_2)Cl_3 + NH_4^+ + Cl^-$$

$$HgCl_2 + 2NH_3 \rightarrow Hg(NH_2)Cl + NH_4^+ + Cl^-$$

$$BCl_3 + 6NH_3 \rightarrow B(NH_2)_3 + 3(NH_4^+ + Cl^-)$$

(dissolving of BCl_3 leads to a solution of strong acid

(NH_4^+), whilst that of BF_3, which is not ammonolysed, leads only to a solution of the weak acid NH_3BF_3):

$$SO_2Cl_2 + 4NH_3 \rightarrow SO_2(NH_2)_2 + 2(NH_4^+ + Cl^-)$$
$$HSO_3Cl + 3NH_3 \rightarrow SO_3NH_2^- + 2NH_4^+ + Cl^-$$
$$RCl + 2NH_3 \rightarrow RNH_2 + NH_4^+ + Cl^-$$
$$RHgCl + NH_3 \rightarrow RH + Hg(NH_2)Cl$$
$$(CH_3CO)_2O + NH_3 \rightarrow CH_3CONH_2 + CH_3CO_2H$$
$$C_2H_5COOR + NH_3 \rightarrow C_2H_5CONH_2 + ROH.$$

In liquid hydrofluoric acid, we have the same taking place with 'fluorolysis' reactions, in which the F^- anions of the solvent take part:

$$SbCl_5 + 6HF \rightarrow SbClF_5^- + H_2F^+ + 4HCl \text{ (at } 0\,°C)$$
$$UO_2F_2 + 6HF \rightarrow UF_6 + 2(H_3O^+ + F^-)$$
$$HSO_3OH + 2HF \rightarrow HSO_3F + H_3O^+ + F^-$$
$$(CH_3CO)_2O + HF \rightarrow CH_3COF + CH_3CO_2H.$$

II.3. Protolysis in Molecular Solvents of Low Dielectric Constant

We have previously seen (Section I.3) that the characteristic feature of molecular solvents of low dielectric constant is the importance of the phenomenon of ionic association (formation of ion-pairs) whereby the concentration of free ions becomes progressively lower when ε falls. This brings with it some changes to the argument in the treatment of the properties of acids and bases in these media, which are fairly frequently encountered.

It is in fact a matter of superposing on the acid-base equilibria proper, such as we have described and applied in Section II.1, the different equilibria of ionic association which may intervene and in which ionic or basic species take part.

To bring out the essential features of the argument and the consequences of ionic association, we will consider more especially the example of a very common solvent with a low dielectric constant, anhydrous acetic acid ($\varepsilon = 6.2$).

PROPERTIES OF ACIDS AND BASES IN ANHYDROUS ACETIC ACID

Acetic acid (AcOH) belongs to the class of amphiprotic solvents; it is characterized by the ionic product:

$$K_S = [AcOH_2^+]\,[AcO^-] \approx 10^{-15}\,mol^2\,l^{-2}.$$

$AcOH_2^+$ represents the solvated proton, the strongest acid existing in this medium. The acetate ion AcO^- is the strongest base. The pH is defined, as we have seen previously, by:

$$pH = -\log(\gamma_{AcOH_2^+}\,[AcOH_2^+])\,. \qquad \text{(II-44)}$$

The pure neutral solvent is characterized by $pH \approx 7.5$.

Kolthoff and his coworkers introduced (starting in 1956) consideration of the effect of ion-pair formation so as to obtain a precise interpretation of the acid and base properties in this solvent (see Kolthoff and

Bruckenstein, 1959). Some essential features of the analysis resulting from this are given below.

State of acids and bases in acetic acid solution

1. Let us first consider a 'strong' acid, perchloric for example. It is practically completely protolyzed by acetic acid (then acting as a base), leading to solvated H^+ cations (that is to say $AcOH_2^+$) and ClO_4^- anions. But due to the low dielectric constant, these two ions of opposite charges will remain associated for the greater part, forming an ion-pair, and there will in fact only be a very small number of free H^+ and ClO_4^- ions in solution. The association constant of the acid, the reciprocal of the ionic dissociation constant, is equal to that of a weak acid: $pK_A = 4.85$.

It therefore appears that the substantial ionic association is responsible for the observed behaviour as a weak acid (from the viewpoint of dissociation into free H^+ ions) of all acids including the strongest, when in a solvent of low dielectric constant. The result is that the lowest pH values which can be observed are considerably greater than those of strong acids in solvents of high dielectric constant; for example a 1 M solution of perchloric acid in acetic acid gives a $pH \approx 2.4$.

2. Let us now consider a 'strong' base, such as sodium acetate. There again, oppositely charged ions Na^+ and AcO^- will remain mostly associated in the form of

ion-pairs; the logarithm of the association constant, identical here to pK_B, has a value of 6.7. Sodium acetate therefore behaves in practice as a weak base; a 1 M solution of this base, for example, gives $pH \approx 10.9$.

In the same manner, ammonia and pyridine are in principle 'strong' bases, that is to say that they are practically completely protolyzed by acetic acid (acting then as an acid), leading to formation of AcO^- and NH_4^+ or pyridinium$^+$ ions. But ammonium and pyridinium acetates form, like sodium acetate, ion-pairs; their pK_B values are thus respectively 6.4 and 6.1.

3. A true 'weak' acid is distinguished from a strong acid like $HClO_4$ by the fact that it is still less dissociated into free ions than a simple ion-pair. For HCl for example, we have $pK_A = 8.55$. The reason for this is that only the protolyzed fraction of the acid obeys the laws of dissociation of ion-pairs, and in this case this fraction is very small.

We can subdivide the equilibrium of protolytic dissociation into two partial equilibria, one corresponding to ionization of the acid by protolysis and leading to the ion-pair state (characterized by the ionization constant K_i), the other corresponding to the dissociation of this ion-pair (characterized by K_{ass}):

$$HCl + AcOH \underset{\substack{\text{ionization} \\ (K_i)}}{\rightleftharpoons} AcOH_2^+Cl^- \underset{\substack{\text{dissociation} \\ (K_{ass})}}{\rightleftharpoons} AcOH_2^+ + Cl^- . \qquad \text{(II-45)}$$

The observed acidity constant pK_A corresponds to the expression:

$$pK_A = -\log \frac{[AcOH_2^+][Cl^-]}{[HCl] + [AcOH_2^+Cl^-]}$$
$$= \log K_{ass} + \log(1 + 1/K_i). \qquad \text{(II-46)}$$

Again we find that when the value of K_i is very large ('strong' acid), pK_A tends towards the value of $\log K_{ass}$.

The same reasoning can be applied for a 'weak' molecular base, that is one only slightly ionized by protolysis (in the form of acetate).

4. In the case of a weak anionic base, the phenomena observed are still a little more complicated. Protolysis of potassium chloride for example takes place from the action of the following equilibria:

$$K^+Cl^- \rightleftharpoons K^+ + Cl^-; \quad AcO^- + K^+ \rightleftharpoons AcOK \qquad \text{(II-47)}$$
$$Cl^- + AcOH \rightleftharpoons HCl + AcO^- \qquad \text{(II-48)}$$

and the calculation of the pH of a solution of this basic salt (Cl^-) will take into account both the equilibrium (II-48) constant of the protolysis proper, and the ionic association constants of the two ion-pairs potassium chloride ($10^{6.9}$) and potassium acetate ($10^{6.15}$).

Principal consequences

1. The extent of the accessible pH range in an amphiprotic solvent of low dielectric constant is therefore less

than the value of the autoprotolysis constant pK_S. In the case of acetic acid ($pK_S \approx 15$), pH can vary only between about 2 and 12, say only over about ten units.

2. We can note that the pH of a solution of a 'strong' acid such as $HClO_4$, is modified by the presence of one of its salts (formed at the time of neutralization); this latter plays in fact the part of a donor of perchlorate ions, the concentration of which affects the dissociation of the ion-pair $AcOH_2^+ClO_4^-$. Thus a 10^{-2} M solution of $HClO_4$ on its own gives pH ≈ 3.4; in presence of $NaClO_4$ at O.5 M concentration ($\log K_{ass} = 5.5$), we find pH ≈ 4.

Further, no truly 'neutral' salt exists, as in aqueous solution. In fact, due to the (small) differences in ionic dissociation, the autoprotolysis equilibrium of the solvent is always slightly displaced by addition of a salt, such as sodium perchlorate (the result of neutralization of perchloric acid by sodium acetate). Having regard to the values of the autoprotolysis constant of AcOH and the dissociation constants of NaOAc (K_B) and $HClO_4$ (K_A), we calculate for this example:

$$\text{pH} \approx \tfrac{1}{2}pK_S + \tfrac{1}{2}(pK_A - pK_B) = 6.6\,. \qquad \text{(II-49)}$$

The solution of $NaClO_4$ is slightly acidic; its pH is independent of the salt concentration.

3. The quantitative laws of pH variation during neutralization of an acid by a base, or vice versa, are different

from those in a solvent of high dielectric constant (see Figure 3, for example, where the titration curve for perchloric acid, although weakly dissociated, nevertheless has a trend similar to that of the titration curves for strong acids in aqueous solution).

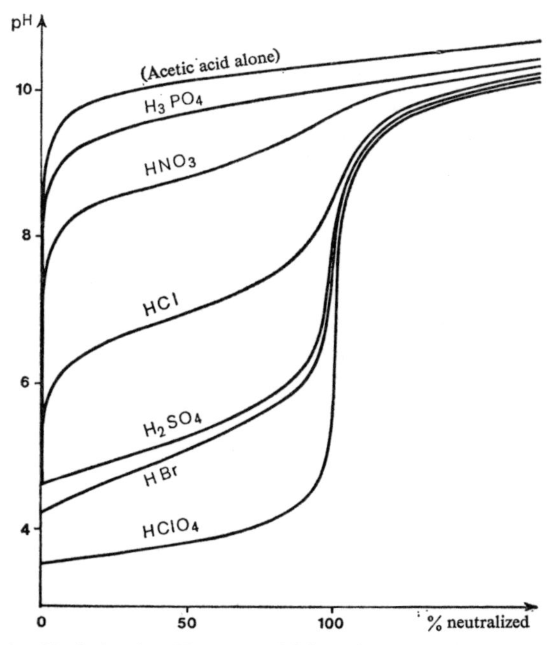

Fig. 3. Variation in pH measured (glass electrode) during neutralization by pyridinium acetate of five inorganic acids 0.02 M in anhydrous acetic acid (From Le Port, 1966).

4. One of the most remarkable consequences of the low ionic dissociation is the fact that coloured indicators no longer indicate the pH in a simple manner as they do in solvents of high dielectric constant, in water particularly.

Let us consider an indicator of which the molecular form I is a weak base. When I reacts with an acid HA, the following equilibria apply:

$$I + HA \rightleftharpoons HI^+A^- \rightleftharpoons HI^+ + A^- . \qquad \text{(II-50)}$$

The protonation of I leads to an acid form with an absorption spectrum different from that of I, but there is no appreciable spectral difference between the free cation HI^+ and this ion associated with another ion of opposite sign. In these conditions, the colour of the indicator is determined, not by the ratio $[HI^+]/[I]$ (which remains connected with the pH, since we always have $pH = pK_I + \log \{[I]/[HI^+]\}$), but by the ratio

$$\frac{[HI^+] + [HI^+A^-]}{[I]} = \frac{[HI^+]}{[I]}\left(1 + \frac{[HI^+A^-]}{[HI^+]}\right).$$

This latter term depends on the degree of association of the acid cation HI^+ with the anion A^-.

When the dielectric constant of the medium is low, the form HI^+A^- predominates. The colour of the indicator, then fixed by the ratio $[HI^+A^-]/[I]$, no longer indicates the pH value of the solution, but that of the concentration of free acid (not combined with groups other than solvent molecules). Actually, from the first of the equilibria (II-50), we see that:

$$\frac{[HI^+A^-]}{[I]} = K[HA]. \tag{II-51}$$

New formulation of acid-base reactions in solution in anhydrous acetic acid

The essential conclusion from this analysis is that reactions between acids and bases (as also is the case with all other reactions) in this solvent take place chiefly between uncharged species, molecules or ionic associations, which are largely predominant over free ionic species (at least in not too dilute solution). The formulation of the reactions that we have mentioned for solvents of high dielectric constant do not therefore represent the principal phenomena, and it is more adequate to refer to another formulation (Trémillon, 1960).

(a) The simpler of the reactions is that between an uncharged acid HA and an uncharged base B; the transfer of a proton leads to the (predominant) form of an ion-pair BH^+A^-. The equilibrium transfer is represented by:

$$B + HA \rightleftharpoons BH^+A^- \tag{II-52}$$

and characterized by the equilibrium constant:

$$K = \frac{[B][HA]}{[BHA]}. \tag{II-53}$$

The reactivities of all bases B with regard to the same acid HA can then be compared according to the values $pK = -\log K$, which increases as the base becomes stronger. Not only the absolute values of pK but also the relative classification of the bases, to a certain extent, depend on the nature of the acid HA under consideration.

In the same way, we can compare the reactivities of all acids HA with regard to the same base B.

(b) The ion-pairs BH^+A^-, associations of an acid cation HB^+ and a basic anion A^-, can react either as acids with a base B′ stronger than B (to take H^+ from it):

$$BH^+A^- + B' \rightleftharpoons B'H^+A^- + B \qquad \text{(II-54)}$$

or as bases with an acid HA′ stronger than HA (to give up H^+ to the A^- anion):

$$BH^+A^- + HA' \rightleftharpoons BH^+A'^- + HA. \qquad \text{(II-55)}$$

By characterizing equilibria of type (II-52) for B and B′ with HA, or for HA and HA′ with B, by constants pK', we can predict that the reactions (II-54) and (II-55) in the left to right direction take place if $pK' \gg pK$.

(c) Anionic bases A^- can exist in solution, associated not only with an acid cation HB^+ (previous case), but also with a non-acid cation, a cation of an alkali metal or of a quaternary ammonium, for example. These pairs of ions can also react with acids HA′:

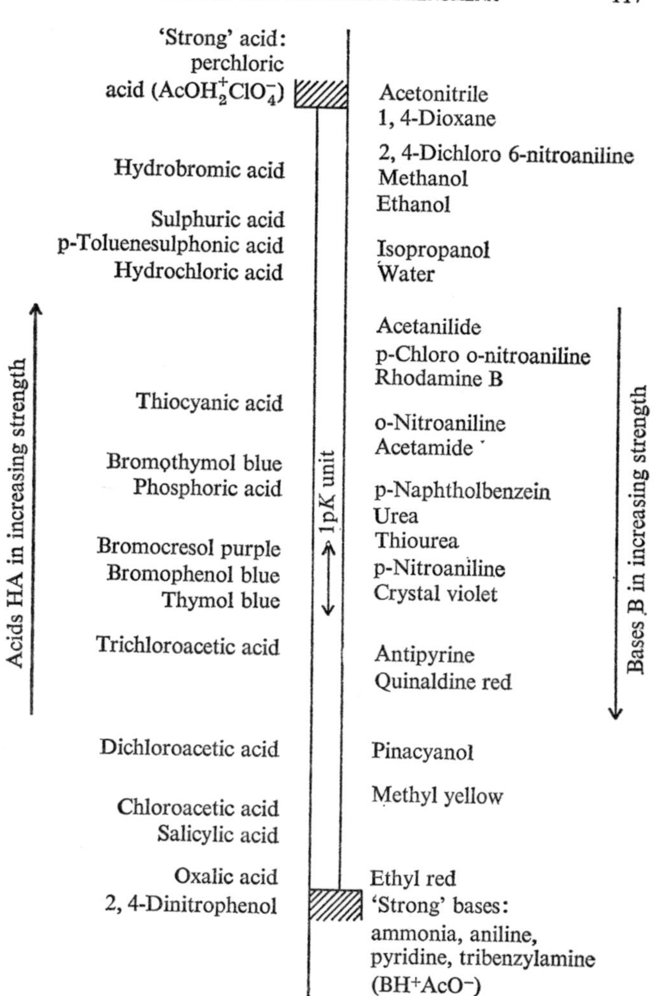

Fig. 4. General acidity scale of molecular acids and bases in anhydrous acetic acid (Trémillon, 1960).

$$M^+A^- + HA' \rightleftharpoons M^+A'^- + HA. \qquad \text{(II-56)}$$

Special equilibrium constants are necessary to predict these reactions.

(d) It is observed that the classification of the strengths of molecular bases or that of the strengths of molecular acids is in fact, roughly speaking, independent of the reference acid or base selected. We can therefore establish a general acidity scale with all the acids HA on one side, and all the bases B on the other side, with the respective distances between an acid HA and a base B corresponding to the value of pK for the reaction between them (Figure 4). This scale enables us to predict semi-quantitatively all acid-base reactions of types (II-52), (II-54) and (II-55), and even, approximately, (II-56). Indicators naturally find a place on this scale (they serve to carry out the measurements of pK), which enables us to predict the possible changes of colour during other acid-base reactions.

Acidity of metallic ions in acetic acid

As in all other amphiprotic solvents, metallic ions and their compounds have an acidic character when in solution in acetic acid (alkaline earth ions, Ag^+, Bi^{3+}, Cd^{2+}, Cu^+, Hg_2^{2+}, Pb^{2+}, Zn^{2+}), etc.

The formation of complexes with the acetate anion AcO^- is responsible for this phenomenon. Under these conditions, a large number of acetates are sufficiently soluble for there to be no precipitation when they are formed (Touller and Trémillon, 1970).

ACID-BASE REACTIONS IN OTHER SOLVENTS OF LOW DIELECTRIC CONSTANT

The formation of ion-pairs is to be taken into account in treatment of properties of acids and bases, in a large number of solvents, since this phenomenon becomes appreciable for $\varepsilon <$ about 40. We can still neglect it up to about $\varepsilon \approx 20$, provided that we are only considering fairly dilute solutions. On the other hand, an exact analysis of acido-basic properties in liquid ammonia ($\varepsilon \approx 20$), isopropanol ($\varepsilon \approx 18$), ethylenediamine ($\varepsilon \approx 14$) or in pyridine ($\varepsilon \approx 12$) requires it to be taken into consideration, without ever being able to neglect ionic dissociation. A treatment similar to that of acetic acid is valid only for solvents of dielectric constant below about 10; this is the case, for example, in trifluoroacetic acid ($\varepsilon = 8.3$, amphiprotic), tetrahydrofuran ($\varepsilon = 7.6$), and 1,4-dioxane ($\varepsilon = 2.2$, protophilic solvents), and quite specially in various organic solvents which can be regarded as practically aprotic: benzene ($\varepsilon = 2.3$), chlorobenzene ($\varepsilon = 5.6$), chloroform ($\varepsilon = 4.8$), carbon tetrachloride ($\varepsilon = 2.2$).

In these latter cases, the phenomena are seen to be still more complicated by the superposition onto the hydrogen ion transfer equilibria (II-52, II-54, II-55 and II-56) of equilibria of association by hydrogen bonding, for example:

$$2HA \rightleftharpoons (HA)_2 \quad \text{(II-57)}$$

$$2BHA \rightleftharpoons (BHA)_2 \quad \text{(II-58)}$$

$$BHA + HA \rightleftharpoons B(HA)_2 \quad \text{(II-59)}$$

(Davis, 1968, Rumeau, 1971).

II.4. Protolysis in Ionized Solvents (Molten Salts)

Experiments have shown that, if the temperature is not too high, certain acids and bases, molecular as well as ionic, can be dissolved in molten salts. We can therefore also carry out acid-base reactions in these media (that is H^+ exchanges between donors and acceptors).

The solvent ions are themselves capable, if they exhibit protogenic and/or protophilic features, of taking part in protolysis equilibria, and we can distinguish, as for molecular solvents, ionized solvents which are amphiprotic, or protophilic only.

The amphiprotic nature can be due to the fact that:

one of the solvent ions is itself amphiprotic, the other being aprotic;

or one of the ions is protogenic, whilst the other is protophilic.

We will give only a small number of examples of each of the different cases.

1. A uniquely protophilic ionized medium in which the

acido-basic properties have been investigated experimentally (at 127 °C, Vedel and Trémillon, 1966) is molten ethylpyridinium bromide (m.p. 114 °C). The ethylpyridinium cation is aprotic, but the Br^- anion is protophilic by nature, and we can expect that a dissolved acid HA would be protolyzed in accordance with the equilibrium:

$$HA + Br^- \rightleftharpoons HBr + A\,. \tag{II-60}$$

Dissolved hydrobromic acid is thus the strongest acid which can exist in this medium; it corresponds to the solvated proton in the other protophilic molecular solvents.

The acid-base couple HA/A is, if protolysis of HA in the molten salt is not complete (that is if HA is not a strong acid), characterized by the acidity constant:

$$K_A = \frac{[HBr]\,[A]}{[HA]} \tag{II-61}$$

(regarding activities and concentrations as equivalent, and taking the activity of the Br^- ion as unity, which implies that the solution is sufficiently dilute).

In order to measure the acidity of such a medium, we define a characteristic quantity which plays the same part as the pH in protophilic molecular solvents; this is $pHBr = -\log[HBr]$. Thus, a mixture of the constituents of the couple HA/A is characterized by the value:

$$pHBr = pK_A + \log\frac{[A]}{[HA]} \tag{II-62}$$

pHBr can be measured like pH (hydrogen electrode).

Different acid-base couples have been observed and their constants pK_A determined: for example, salicylic acid 5.8, formic acid 9.2, acetic acid 10.2, benzoic acid 8.7, phosphoric acid 7.8, pyridine 5.3, ammonia 10.2. Acid-base reactions have thus been predicted and carried out.

2. Another medium similar to the previous one, but with a protogenic cation which gives it an amphiprotic nature, was made the subject of a similar investigation at the same temperature (Vedel, 1968); this medium is molten ethylammonium chloride (m.p. 108 °C). In this case, the strongest acid, corresponding to the solvated proton, is dissolved hydrochloric acid. At the other extreme, ethylamine, the conjugate base of the ethylammonium cation, is the strongest base. Dissolved bases are therefore protolyzed by the solvent cation; many common bases are strong: oxide, hydroxide, acetate, fluoride ions and ammonia, for example.

The pHCl scale measuring the extent of the accessible acidity range is limited to about 8 units because of this protogenic character, which corresponds to an autoprotolysis equilibrium of the molten salt and to its constant:

$$C_2H_5NH_3^+ + Cl^- \rightleftharpoons C_2H_5NH_2 + HCl \qquad \text{(II-63)}$$

$$K_S = [C_2H_5NH_2]\,[HCl] = 10^{-8.1}\ \text{mol}^2\,\text{kg}^{-2} \qquad \text{(II-64)}$$

(the activities of the solvent ions, not varying appreciably, are conventionally taken as equal to unity).

3. The most important ionized medium of amphiprotic nature is without any doubt constituted by molten alkali hydroxides. It is the hydroxide ions, simultaneously protophilic and protogenic, which, in this case, bring to them only the amphiprotic nature. Their autoprotolysis equilibrium is expressed by:

$$2OH^- \rightleftharpoons H_2O + O^{2-} \tag{II-65}$$

(this is the equilibrium of dehydration of hydroxides into oxides). Following the usual reasoning, dissolved water is the strongest acid which can exist in molten hydroxides; it corresponds to the solvated proton. At the other extreme, the oxide ion O^{2-} is the strongest base. Thus, hydroxides exert a double protolytic effect on the acids and on the bases that can be dissolved in them. Most of the common acids which are still stable at the utilization temperatures of molten hydroxides are strong, such as:

$$HSO_4^- + OH^- \rightarrow H_2O + SO_4^{2-}$$

$$NH_4^+ + OH^- \rightarrow H_2O + NH_3$$

$$C_6H_5OH + OH^- \rightarrow H_2O + C_6H_5O^- .$$

Ammonia, which no longer has any basic property, since $NH_4{}^+$ is a strong acid, behaves as a weak acid, the weak conjugate base of which is the amide anion:

$$NH_3 + OH^- \rightleftharpoons H_2O + NH_2^- . \qquad \text{(II-66)}$$

But whilst amide is a weak base protolyzed to a negligible degree ($NH_2^- + OH^- \rightleftharpoons O^{2-} + NH_3$), hydride H^- is a strong base:

$$H^- + OH^- \rightleftharpoons O^{2-} + H_2 .$$

Measurement of acidity in molten hydroxides is carried out by means of the quantity

$$pH_2O = -\log [H_2O], \qquad \text{(II-67)}$$

which plays a part similar to that of pH.

The lowest pH_2O values (the most acidic media) correspond to the highest concentrations of dissolved water, these latter being dependent on the temperature and on the partial vapour pressure of water in equilibrium with the solution. The highest values of pH_2O correspond to the highest concentrations of dissolved alkali oxide, linked with pH_2O by the value of the autoprotolysis constant:

$$K_S = [H_2O]\,[O^{2-}] \qquad \text{(II-68)}$$

(the activity of the OH^- ions is taken as unity).

The experimental investigation carried out at 227 °C on the eutectic mixture NaOH+KOH (m.p. 170 °C) led to the value $K_S \approx 10^{-11.6}$ mol^2 l^{-2} (Goret and Trémillon, 1966). At this temperature, under very small partial pressures, the solubility of water can reach values approaching 10 M, that is $pH_2O = -1$. At the other extreme, the solubility of alkali oxides at this temperature is close to 0.1 M, that is $pH_2O = 10.6$. The extent of the acidity scale available is about 11.5 units under these conditions.

The molten eutectic, completely pure, neutral, has $pH_2O = 5.8$. We are therefore dealing with an acidic medium if $pH_2O < 5.8$, and a basic medium if $pH_2O > 5.8$.

Couples of weak acid with weak base are characterized in the usual manner by means of an acidity constant; for example, for NH_3/NH_2^- (equilibrium II-66):

$$K_A = \frac{[H_2O]\,[NH_2^-]}{[NH_3]} \approx 10^{-3}\ \text{mol l}^{-1}. \qquad \text{(II-69)}$$

Since the medium is amphiprotic, we can expect that the dissolved metallic ions (in the state of hydroxides) will also behave as acids. In fact, their accepting strength of the strong basic ion O^{2-} determines the displacement of the dissociation equilibrium of the hydroxide anions:

$$2OH^- \rightleftharpoons H_2O + O^{2-}$$

leading to liberation of water molecules, and therefore to acidification of the solution. The acid character of metallic cations is thus shown by the lowering in the pH_2O value when they are dissolved.

In the same way that most metallic hydroxides are only slightly soluble in water and are precipitated when formed, the solubility of metallic oxides in molten alkali hyroxides is usually very low. In consequence, the acid-base equilibrium of a metallic cation can for example be written:

$$M^{n+} + nOH^- \rightleftharpoons MO_{n/2}(s) + \tfrac{1}{2}nH_2O \qquad \text{(II-70)}$$

an equilibrium which is equivalent, in aqueous solution, to:

$$M^{n+} + 2nH_2O \rightleftharpoons M(OH)_n(s) + nH_3O^+ . \qquad \text{(II-71)}$$

A metallic oxide can thus be precipitated when an acidic solution, that is a hydrated solution, of the corresponding hydroxide is neutralized (by addition of alkali oxide or by elimination of water in the form of steam). In the opposite direction, an insoluble metallic oxide can be redissolved by acidification (dissolving of water, brought in either in the form of steam, or by introduction of pellets of hydrated sodium or potassium hydroxide). Certain oxides are amphoteric and can also be redissolved by reduction of the acidity or by increase of the basicity, due to the formation of higher anionic oxo-complexes. All in all, precipitation of a metallic oxide in molten hydroxides takes place in a general manner within a certain acidity range, the limits of which can be calculated exactly from the value of the initial concentration of the cation and those of its acidity constants (or from the solubility product of the oxide).

As we usually do in the case of solubility of hydroxides in aqueous solution, the phenomena can be graphically represented by means of diagrams of apparent solubility of the metallic oxides as functions of pH_2O (equilibrium

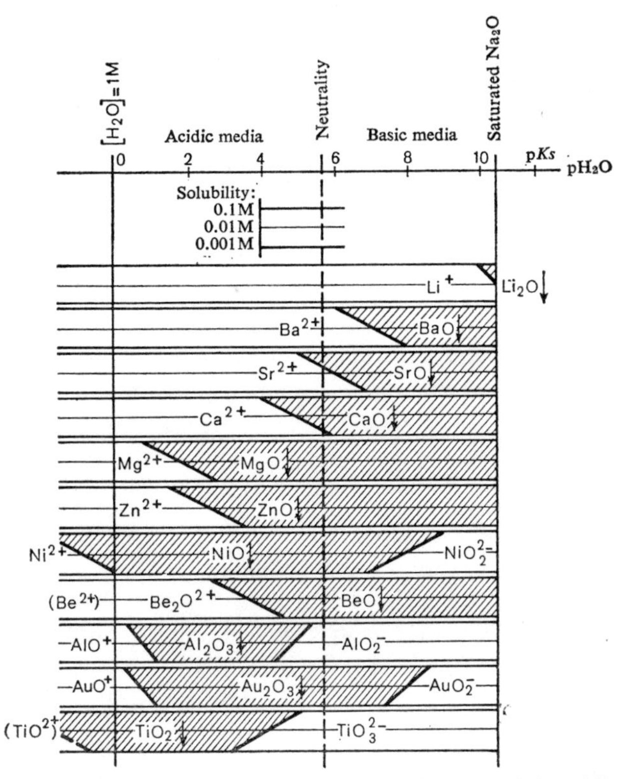

Fig. 5. Diagram of apparent solubility as a function of the acidity (pH_2O) of some metallic oxides in the eutectic mixture of molten NaOH + KOH at 227°C (Goret and Trémillon, 1966).

diagrams). Figure 5 shows an outline sketch of these diagrams obtained by experiment for a certain number of ions in the NaOH+KOH eutectic at 227 °C.

It is evident that these results are important for explanation or prediction of the dissolving power of molten hydroxides with regard to solid materials made of metallic oxides. From the diagrams it is possible to classify oxides according to their ability to dissolve or be precipitated in acidic, basic or neutral media. Thus, TiO_2 and ZrO_2 are insoluble in a strongly acidic medium, Al_2O_3 is only slightly soluble (in the more acidic form AlO^+; it is to be noted that the diaspore Al_2O_3,H_2O corresponds to $2AlO^+,2OH^-$; this is therefore the directly soluble grouping). In a neutral medium (dehydrated), ZrO_2, ZnO, BeO, MgO, etc., are insoluble, whilst TiO_2 dissolves in the form of titanate, Al_2O_3 is again slightly soluble but in the form of aluminate AlO_2^-. In a basic medium, alkaline earth oxides become insoluble, whilst alumina and TiO_2 are very soluble. Oxides which are very acidic, SiO_2, V_2O_5, etc., dissolve over the whole acidity range, behaving as strong acids; SiO_2 dissolves, liberating two molecules of water, which corresponds to formation of orthosilicate SiO_4^{4-}. VO_3^- and PO_3^- are also strong acids.

In conclusion, it appears that there are no grounds for making a distinction, with regard to methods of reasoning, between ionized and molecular solvents. The examples given are sufficient to show that protolysis follows the same rules and obeys the same general laws in the two cases. This fact certainly opens interesting horizons for chemical applications at high temperatures. Further, it justifies the transfer of reasoning with regard to these media, to categories of properties other than protolysis, as will be shown during the following chapter.

CHAPTER III

OTHER ION EXCHANGE SYSTEMS WITH PARTICIPATION OF SOLVENT

III.1. General Concepts of Acidity (Non-Protonic) in Aprotonic Solvents

The success of the acid-base theories of Arrhenius and of Brønsted–Lowry has led to several attempts at rationalization of other types of chemical reaction in solution, with the aim of providing a system of reasoning applicable to various properties and to various media, and to facilitate a more effective application of these latter. Certain of these efforts undeniably constitute at least some progress towards harmonization of the whole subject, so vast and so diversified, chemistry in solution.

A fundamental concept, which appears as the generalization of Brønsted–Lowry, is the idea of the acceptor-donor couple of a chemical 'particle' other than the hydrogen ion, in particular a simple ion such as halide, oxide, sulphide, etc.*

* We will consider separately the electron acceptor-donor couples or oxidizing-reducing systems, which form the subject of a special treatment (Chapter IV) and which however present a remarkable similarity to ionic exchange systems.

By giving a treatment particularly applicable to one simple determined ion – as was done in the preceding chapter for the H^+ ion for protolysis phenomena – we will distinguish the compounds which are donors of this ion and their conjugate acceptors, constituting the family of acceptor-donor couples of the ion under consideration. Ion-exchange reactions between a donor of one couple and an acceptor of another couple can then be considered. If the reactions correspond to chemical equilibria, they will be predictable starting from the characteristic equilibrium constants of the various couples.

As is the case for protolysis, the solvent can itself intervene in these phenomena, either as acceptor of the ionic particle under consideration, or also as donor (it is then amphoteric).

This general conception is that of 'ionotropy' (Gutmann and Lindqvist, 1954). It comprises as a special case the 'prototropy' of Brønsted–Lowry, each type of exchange reaction being treated in a manner similar to the exchange reactions of H^+. It has in particular been the subject of important applications in molten salts, with the concept of 'oxoacidity'.

Concept of acidity (oxoacidity) in molten salts according to the definition of Lux-Flood

Lux (1939) and then Flood and co-workers (1947) suggested a new definition of acids and bases, valid for media constituted from molten salts, on the exchange of the oxide ion O^{2-} (oxidotropy), acid-base couples being acceptor-donors of this ion. The justification of this new definition appears to be as much practical as theoretical, since we note that classical acids (protic compounds) mostly cease to exist in solution at high temperatures, whilst on the contrary compounds of the ion O^{2-} take on an increased importance, having regard to their stability, their number and natural abundance, and their relation with the oxidation-reduction phenomena where oxygen is concerned; in addition, they give rise to generally well established exchange equilibria (of O^{2-}), in contrast with their most common behaviour at ordinary temperature. In order to avoid possible confusion with normal acids and bases, it has been suggested that acceptors of O^{2-} (metallic cations, 'anhydrides' such as CO_2, SO_2, SO_3, P_2O_5, various other oxides such as SiO_2, Cr_2O_3, Al_2O_3 and of course H_2O) should be designated as 'oxoacids', and donors of O^{2-} (oxo-anions such as CO_3^{2-}, PO_4^{3-}, SO_4^{2-}, SiO_4^{4-}, BO_4^{3-}, AlO_2^-, CrO_4^{2-}, dissociable metallic oxides, and of course water which is amphoteric) should be designated as 'oxobases'.

Oxoacid-oxobase couples are defined by the general relation:

$$\underset{\text{(donor)}}{\text{oxobase}} = \underset{\text{(acceptor)}}{\text{oxoacid}} + O^{2-} . \qquad \text{(III-1)}$$

It is to be noted that, having regard to the opposition in charge between H^+ and O^{2-}, here it is the acceptor and not the donor which has the oxoacid function. The following are examples of couples:

$$
\begin{array}{lll}
MgO & = Mg^{2+} & + O^{2-} \\
\left\{\begin{array}{l} 2AlO_2^- \\ Al_2O_3 \\ AlO^+ \end{array}\right. & \begin{array}{l} = Al_2O_3 \\ = 2AlO^+ \\ = Al^{3+} \end{array} & \begin{array}{l} + O^{2-} \\ + O^{2-} \\ + O^{2-} \end{array} \\
\left\{\begin{array}{l} SiO_4^{4-} \\ SiO_3^{2-} \\ SiO_2 \\ SiO^{2+} \end{array}\right. & \begin{array}{l} = SiO_3^{2-} \\ = SiO_2 \\ = SiO^{2+} \\ = Si^{4+} \end{array} & \begin{array}{l} + O^{2-} \\ + O^{2-} \\ + O^{2-} \\ + O^{2-} \end{array} \\
UO_3 & = UO_2^{2+} & + O^{2-} \\
CO_3^{2-} & = CO_2 & + O^{2-} \\
\left\{\begin{array}{l} 2SO_4^{2-} \\ S_2O_7^{2-} \end{array}\right. & \begin{array}{l} = S_2O_7^{2-} \\ = 2SO_3 \end{array} & \begin{array}{l} + O^{2-} \\ + O^{2-} \end{array} \\
\left\{\begin{array}{l} 2CrO_4^{2-} \\ Cr_2O_7^{2-} \\ CrO_3 \end{array}\right. & \begin{array}{l} = Cr_2O_7^{2-} \\ = 2CrO_3 \\ = CrO_2^{2+} \end{array} & \begin{array}{l} + O^{2-} \\ + O^{2-} \\ + O^{2-} \end{array} \\
\left\{\begin{array}{l} H_2O \\ 2OH^- \end{array}\right. & \begin{array}{l} = 2H^+ \\ = H_2O \end{array} & \begin{array}{l} + O^{2-} \\ + O^{2-}. \end{array}
\end{array}
$$

In order to measure the oxoacidity in a molten salt where free O^{2-} ions can exist, recourse is made to a characteristic quantity similar to the pH and playing the same part:

$$pO^{2-} = -\log a_{O^{2-}}. \qquad \text{(III-2)}$$

At a sufficient dilution, the activity $a_{O^{2-}}$ can be replaced by the concentration $[O^{2-}]$. In contrast to pH, value of which becomes smaller as the acidity increases (since H^+ is the 'gene' of acidity), the pO^{2-} value increases with oxoacidity (since O^{2-} is the 'gene' of oxobasicity).

SOLVOACIDITY

The ideas of acidity and of oxoacidity, based on the consideration of the two most important ions in nature, H^+ and O^{2-}, are independent of the solvent in which they are applied. In contrast, a generalized concept of acidity in non-aqueous solvents, clearly inspired by Arrhenius' theory on aqueous solutions, has been introduced by linking the definition of acids and bases with a property characteristic of the solvent: the self-ionization of its molecules (Germann, 1925; Cady and Elsey, 1928; Franklin, 1935; Gutmann, 1952). The nature of the solvoacid-solvobase couples is determined by the ion exchanged between two molecules of solvent to produce this self-ionization (more generally self-dissociation, to include the case of ionized solvents).

In fact, a fairly large number of molecular solvents exhibit a small degree of self-ionization which takes place by means of various processes. In the previous chapter, we considered the case of ionization by autoprotolysis (amphiprotic solvents), that is by transfer of a H^+ ion between two amphoteric solvent molecules. But aprotonic solvents can give rise to the same type of phenomenon by means of other ionic species being transferred: Cl^-, Br^-, F^-, O^{2-}, etc. This is the case, for example, with phosphorus or selenium oxychlorides, nitrosyl chloride, antimony trichloride (molten), bromine trichloride, nitrogen dioxide, etc, in which the following self-ionization equilibria have been postulated:

$$
\begin{aligned}
2POCl_3 &\rightleftarrows POCl_2^+ + Cl^-, POCl_3 \\
2SeOCl_2 &\rightleftarrows SeOCl^+ + Cl^-, SeOCl_2 \\
2NOCl &\rightleftarrows NO^+ + Cl^-, NOCl \\
2SbCl_3 &\rightleftarrows SbCl_2^+ + SbCl_4^- \\
2HgBr_2 &\rightleftarrows HgBr^+ + HgBr_3^- \\
2BrF_3 &\rightleftarrows BrF_2^+ + BrF_4^- \\
N_2O_4 (\rightleftarrows 2NO_2) &\rightleftarrows NO^+ + NO_3^- .
\end{aligned}
$$

All these equilibria lead only to very low degrees of self-ionization; the solvents therefore remain essentially molecular – like water, alcohols, ammonia, acetic acid, etc. – despite autoprotolysis.

The self-ionization equilibrium of a solvent of this type is known as the 'solvent-system'; it leads to the appearance of two ions, a cation and an anion, characteristic of the solvent under consideration. From this comes the definition of 'solvoacids' and 'solvobases' (Cady and Elsey, 1928):

(a) A solvoacid is a solute which yields the characteristic solvent cation, either by direct dissociation (due to solvolysis) or by reaction with the solvent molecules (then acting as an acceptor of the characteristic anion).

(b) A solvobase is a solute which yields the characteristic solvent anion, either by direct dissociation or by reaction with the solvent molecules (then acting as an acceptor of the characteristic cation).

For aqueous solutions, such a definition is in as good agreement with the older Arrhenius theory as with that of Brønsted–Lowry. It also leads to a re-statement for other amphiprotic solvents (HS) of the definition of classical acids and bases: respectively donors (by direct dissociation) of H_2S^+ and therefore acceptors of S^-, and acceptors of H^+ (that is of H_2S^+) and therefore donors of S^-. Naturally in this definition, acceptors of S^- (metallic cations, acid anhydrides, etc) appear as solvoacids.

But, for aprotonic solvents, the definition leads to phenomena being considered as acidity when they are not protolysis phenomena. To avoid all confusion, it is then necessary to specify, as is done for oxoacidity, the nature of the solvoacidity determined by the solvent-system according to the ionic particle transferred. Thus, the term 'chloroacidity' will be used for solvoacidity in the solvents $POCl_3$, $SeOCl_2$, NOCl, $SbCl_3$, etc, where the acid-base couples (more precisely chloroacid-chlorobase couples) in all these solvents are acceptor-donor couples of the Cl^- ion.

The solvents in which the solvoacidity theory applies in its non-protonic form are molecular solvents which are nevertheless very active chemically, such as molten salts, in which protonic exchanges do not take place, or have only a limited interest; chemical properties in these media are very different from properties in aqueous solution or in the solvents closest to water. It does not therefore seem that use of terms recalling the more common properties can lead to any real confusion.

Nevertheless, we must not exaggerate the importance of this theory. In the first place, it extends the notion of acidity only to a limited number of solvents, not in common use, except perhaps in the range of molten salts (above all with the notion of oxoacidity; see later). In the second place, it is only a unification of reasoning and it does not make possible any quantitative comparisons between the different sorts of solvoacidity.

Other generalized definitions of acids and bases

Several other definitions of acids and bases which have been suggested do not greatly differ from the preceding definitions and have not received a very great acceptance (Ussanovich, 1939; Ebert and Konopik, 1949).

Although it does not form part of the ionic theories to which this discussion is limited, it is impossible to omit the concept of acidity put forward by Lewis (1938), which forms an 'electronic' theory of acids and bases (cf. Luder and Zuffanti, 1946).

Lewis designated as an 'acid' any compound which accepts an electron-pair (such as BF_3, $AlCl_3$, $SbCl_5$, SO_3, H^+, I_2, etc.) and as a 'base' any compound which is a donor of an electron-pair (such as amines R_3N, nitriles RCN, ethers R_2O, H_2O, $POCl_3$, the anions CN^-, SCN^-, O^{2-}, etc.). This classification enables us to predict the possibilities of a reaction between an 'acid' (acceptor) and a 'base' (donor) to produce a coordination compound ('complex').* In those circumstances, it is an addition and not an exchange reaction as in the ionic theories; but exchange reactions can flow from the competition between two acceptors for the same donor, or between two donors for the same acceptor. Thus, a classic acid-base reaction $HA+B \rightleftharpoons A+HB$ is the result of competition between the two bases :A and :B for the acceptor H^+. This is the analogue for example of the exchange:

$$(AlCl_3 \leftarrow A) + :B \rightleftharpoons (AlCl_3 \leftarrow B) + :A$$

which is the result of competition between the two bases :A and :B for the acceptor $AlCl_3$.

The terms 'Lewis acids' and 'Lewis bases' are in frequent use in the domain of organic chemistry. Specialists in coordination chemistry, who make the most systematic use of Lewis' theory, prefer to use the terms

* In the most general manner, we designate by 'complex' any chemical combination which unites an acceptor and a donor, either molecular or ionic, and by extension any compound which can be divided into two parts, acceptor and donor. Metallic complexes are those where the acceptor is a metallic cation M^{n+}.

'acceptor' and 'donor' so as to avoid confusion.

Our interest in Lewis' general theory is that it throws light on the system of chemical reactions in solution (with the exception of oxidation-reduction reactions), which all take place in accordance with the same coordination mechanism. But it brings nothing new to quantitative prediction of these reactions, which can be only effected by families of reactions.

In connection with the properties of non-aqueous solvents, an interesting application has recently been mentioned for interpretation and classification of solvation phenomena, by reference to the donor role of the solvent (Drago and Meek, 1961). Donor solvents have been classified quantitatively according to the values of their reaction enthalpies, in dichloroethane, with a common acceptor, antimony pentachloride ('donor number' DN, Gutmann *et al.*, 1966–1968). The weakest donor solvents, for example, other than hydrocarbons and their halogenated derivatives ($DN \approx 0$) are: nitromethane, $DN = 2.7$; nitrobenzene, $DN = 4.4$; acetic anhydride, 10.5, acetone, 17.0; water, 18.0. Stronger donors than water are diethyl ether, $DN = 19.2$; THF, 20.0; DMF, 26.6, DMSO, 29.8. The strongest donor solvents are pyridine, 33.1, and HMPA, $DN = 38.8$.

III.2. Examples of Solvoacidity in Aprotonic Solvents

We will now describe some examples of aprotonic solvoacidity, stressing in particular the similarity of the argument with that relating to protonic acidity in amphiprotic solvents, which is an essential feature of this theory.

CHLOROACIDITY IN SOLVENTS WHICH ARE DONORS OF THE CHLORIDE ION

1. *Molecular solvents.* Let us consider a solvent the molecule of which – symbolized by $S\mathrm{Cl}$ – can play the part of chloride ion donor, and which from this fact can be the seat of a self-ionization equilibrium represented by:

$$S\mathrm{Cl} \rightleftharpoons S^+ + \mathrm{Cl}^- \quad \text{or} \quad 2S\mathrm{Cl} \rightleftharpoons S^+ + S\mathrm{Cl}_2^- . \tag{III-3}$$

This is the solvent-system, showing the existence of two characteristic ions (in the solvated state), one being the Cl^- anion solvated (represented equally well by Cl^- as by $S\mathrm{Cl}^-$). This equilibrium is characterized by the constant:

$$K = \frac{(a_{S^+})\,(a_{\mathrm{Cl}^-})}{a_{S\mathrm{Cl}}} . \tag{III-4}$$

If ionization remains very low, the activity of $S\mathrm{Cl}$ may be taken as practically equal to unity (this will be the case in any dilute solution); the self-ionization constant then takes the form of a product of activities, the ionic product of the solvent:

$$K_S = (a_{S^+})\,(a_{\mathrm{Cl}^-}) . \tag{III-5}$$

For a small ionic strength, where we can regard activity as equal to concentration:

$$[S^+]\,[Cl^-] \approx K_S. \tag{III-6}$$

The ionization equilibrium is affected by dissolution of substances which are donors or acceptors of S^+ or of Cl^-, that is to say chloroacids and chlorobases. The first of these are made up of acceptors of Cl^-: metallic cations, H^+ (perchloric acid, for example), $SbCl_5$, $FeCl_3$, $AlCl_3$, $SnCl_4$, BCl_3, SO_3, etc. Sulphur trioxide for example is a chloroacid by formation of the chlorosulphate anion SO_3Cl^-:

$$SO_3 + SCl \rightleftharpoons SO_3Cl^- + S^+.$$

Chlorobases are the donors of Cl^-; the strongest are chlorides of ammoniums (quaternary), the most dissociated (completely so if the dielectric constant of the solvent is sufficiently high); chlorocomplexes, incompletely and sometimes very slightly dissociated, are weak chlorobases, together with acetyl or benzoyl chlorides. Compounds which are acceptors of S^+, such as amines, pyridine, quinoline, HMPA, DMF, DMSO, acetonitrile ('donors' in the sense of Lewis' definition, hence still basic), increase the dissociation of SCl molecules and therefore behave as more or less strong chlorobases according to their strengths of acceptance of S^+. At the same time, there exist some amphoteric compounds ($FeCl_3$, $ZrCl_4$, for example).

Depending on their degree of ionization, which depends both on the nature of the solute and that of the solvent, we may be dealing with: strong chloroacids – quantitative donors of S^+, their conjugate bases then have no basic powers in the solvent concerned – , strong chlorobases – quantitative donors of Cl^- (solvated), the conjugate acids having no acidic powers – , and couples of weak chloroacid and weak chlorobase, each one characterized by the value of an acidity constant (or a basicity constant), similar to the acidity constants (or basicity constants) of the HA/A couples in amphiprotic solvents:

$$K_\mathrm{A} = \frac{[\text{chlorobase}]\,[S^+]}{[\text{chloroacid}]} \quad \text{or} \quad K_\mathrm{B} = \frac{[\text{chloroacid}]\,[\mathrm{Cl}^-]}{[\text{chlorobase}]} \tag{III-7}$$

(where $\mathrm{p}K_\mathrm{A}+\mathrm{p}K_\mathrm{B}=\mathrm{p}K_S$).

Neutralization reactions between chloroacids and chlorobases can be carried out, with quantitative characteristics similar to those of the usual acid-base reactions: strong chloroacid+strong chlorobase:

$$S^+ + \mathrm{Cl}^- \rightarrow S\mathrm{Cl} \tag{III-8}$$

weak chloroacid+strong chlorobase:

$$\mathrm{A} + \mathrm{Cl} \rightarrow \mathrm{ACl} \tag{III-9}$$

weak chlorobase+strong chloroacid:

$$\mathrm{B} + S^+ \rightarrow \mathrm{B}S\,. \tag{III-10}$$

Measurement of chloroacidity in these media is carried out by means of the characteristic quantity:

$$pCl^- = -\log a_{Cl^-} \approx -\log [Cl^-]. \qquad \text{(III-11)}$$

The concentration of the solvated Cl^- ions varies in fact with the nature and the concentration of the dissolved chloroacids and chlorobases.

We will obtain $pCl^- \approx 0$ for a solution of strong chlorobase (Cl^-) at unit concentration, $pCl^- \approx pK_S$ for a solution of strong chloroacid at unit concentration. The extent of the scale of pCl^- accessible is therefore measured by the value of pK_S.

In the pure solvent, exactly neutral, we have $pCl^- = \frac{1}{2}pK_S$.

The variation in pCl^- enables us to characterize the reactions between chloroacids and chlorobases. It is experimentally measurable, by means for example of a chlorine electrode.

Overall, the phenomena of chloroacidity therefore show the same quantitative features as those of the protolysis phenomena in amphiprotic solvents (with this difference that pCl^-, which plays the part of pH in protolysis, increases with chloroacidity).

In selenium oxychloride ($SeOCl_2$), ($\varepsilon = 49$), the value of the ionic product is $K_S = 10^{9.7}$ mol^2 l^{-2}; the extent of the pCl^- scale is thus about ten units. Tetraethylammonium, potassium and triphenylmethyl chlorides, pyridine, for example, are strong chlorobases in this

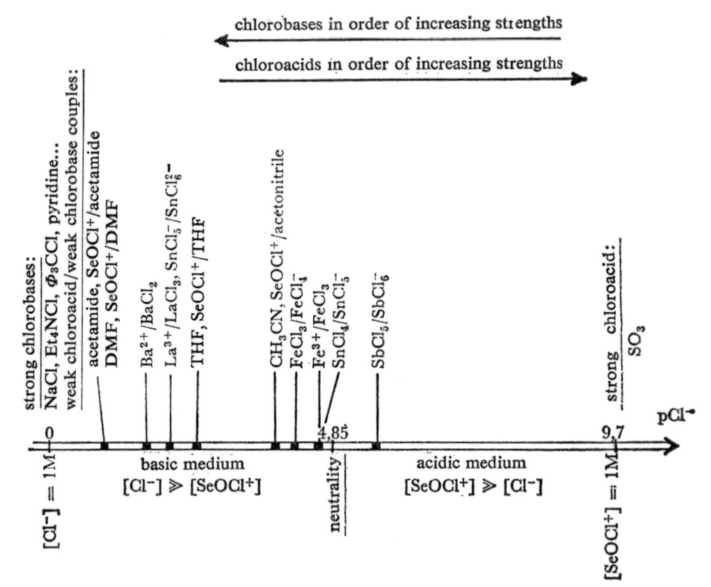

Fig. 6. Solvoacidity in selenium oxychloride.
(a) pCl^- scale; values given by buffer mixtures of weak chloroacid and weak chlorobase in proportions 1/1.

medium; sulphur trioxide is a strong chloroacid (it gives the same chloroacidity, at the same concentration, as $SeOCl^+$ prepared electrochemically); $SbCl_5$ is a weak chloroacid, $pK_A=4.1$, etc.

Figure 6 shows a diagram of these properties, showing various chloroacid-chlorobase couples on a pCl^- scale with a position corresponding to their characteristic value of pK_A. The variations in pCl^- during some

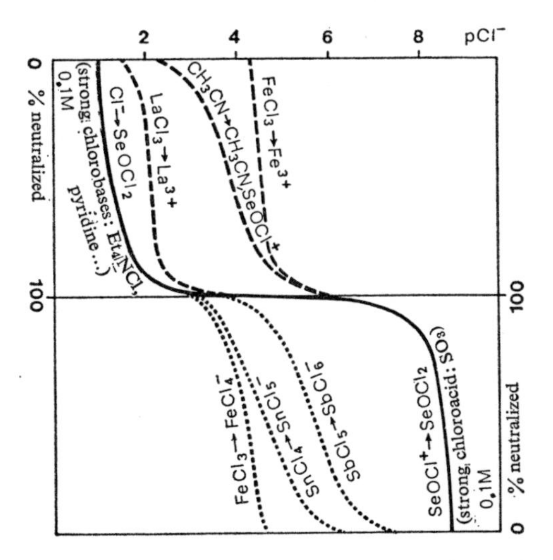

(b) Variations in pCl⁻ (measured by means of a chlorine electrode) during the neutralization reactions of chlorobases by $SeOCl^+$ (– – – –) and of chloroacids by Cl^- (- - - - -) (Devynck and Trémillon, 1969–1971).

neutralization reactions are also shown; it is seen that the curves are quite similar to the variations in pH during acid-base reactions in an amphiprotic solvent (Devynck and Trémillon, 1969–1971).

In phosphorus oxychloride $POCl_3$, the pCl^- scale is appreciably longer, since pK_S is of the order of 19 units. Another important difference with the previous chlorinated solvent is that the dielectric constant is low, $\varepsilon=14$, leading to substantial formation of ion-pairs. All chlorobases and all chloroacids thus behave like weak chlorobases and weak chloroacids; it also appears that ionization (independently of ionic dissociation) of $SbCl_5$ or of SO_3 when solvated is not substantial (Gutmann and Baaz, 1959–1961; Devynck, 1971).

Sulphuryl, thionyl and nitrosyl chlorides, SO_2Cl_2, $SOCl_2$ and NOCl respectively, phosgene $COCl_2$, acetyl and benzoyl chlorides, are also used as solvents and in this respect show considerable similarity with the two previous solvents.

Molten antimony trichloride (m.p. 73 °C, $\varepsilon = 32$ at 100 °C), also constitutes a medium belonging to the same family of solvents which are donors of Cl^-. Its self-ionization system is as follows:

$$2SbCl_3 \rightleftharpoons SbCl_2^+ + SbCl_4^- \qquad \text{(III-12)}$$

($SbCl_4^-$ corresponds to the solvated Cl^- ion). At 100 °C, the ionic product has a value of:

$$K_S = [SbCl_2^+]\,[SbCl_4^-] \approx 10^{-7.8}\ \text{mol}^2\ \text{l}^{-2}.$$

This is therefore a molecular solvent, hardly more ionized than selenium oxychloride at ordinary temperatures.

In this solvent, it has been observed that KCl and triphenylmethyl chloride are strong chlorobases, but that benzoyl chloride is a weak chlorobase; $AlCl_3$ ($\rightarrow SbCl_2^+ + {}$ $+ AlCl_4^-$), and $GaCl_3$ are strong chloroacids, but $TaCl_5$ is a weak chloroacid ($TaCl_5 + SbCl_3 \rightleftharpoons TaCl_6^- + SbCl_2^+$, $pK_A = 4.5$), etc (Texier, 1968).

Another solvent very close to this latter, but liquid at ordinary temperatures, is arsenic trichloride (m.p.—18 °C, $\varepsilon = 12.5$ at 20 °C).

2. *Ionized solvents.* The equimolecular mixture of aluminium chloride and sodium chloride forms the definite compound $NaAlCl_4$ (sodium tetrachloroaluminate, m.p. 151 °C), which in the molten state is completely ionized into Na^+ cations and chlorocomplex anions $AlCl_4^-$. These latter, being donors of the Cl^- ion, also undergo a weak dissociation in accordance with the equilibrium:

$$2AlCl_4^- \rightleftharpoons Al_2Cl_7^- + Cl^- . \tag{III-13}$$

To the extent that dissociation remains sufficiently weak for us to be able to assume that the $AlCl_4^-$ activity remains very nearly unity, this equilibrium is characterized at constant temperature by a constant product of activities, similar to the constants K_S of the molecular solvents previously mentioned:

$$K_S = [Al_2Cl_7^-]\,[Cl^-] . \tag{III-14}$$

According to the definition of solvoacidity, donors of Cl^- are chlorobases as before. Their dissolution provides a medium richer in free Cl^- ions than the exactly equimolecular mixture $AlCl_3 + NaCl$, which corresponds to neutrality ($[Cl^-] = [Al_2Cl_7^-] = K_S^{1/2}$). In particular, an excess of sodium chloride, compared with the equimolecular mixture, provides a strong basic medium. The O^{2-} or F^- ions, which are acceptors of aluminium (III) cations, also behave like chlorobases in molten tetrachloroaluminate, since they cause solvolysis reactions which liberate Cl^- ions, such as:

$$O^{2-} + 2AlCl_4^- \rightarrow Al_2OCl_5^- + 3Cl^- . \qquad \text{(III-15)}$$

In the other direction, the strongest chloroacid which can exist in this solvent is $Al_2Cl_7^-$; it is obtained by addition of an excess of aluminium chloride to the equimolar mixture of $AlCl_3$ and NaCl ($AlCl_3 + AlCl_4^- \rightarrow Al_2Cl_7^-$). Other metallic cations and their compounds are also chloroacids, more or less strong. The mercurous ion for example is a weak chloroacid, assuming the form Hg_2^{2+} in an acidic medium, and being transformed into Hg_2Cl_2 in a basic medium; the following equilibrium is set up:

$$Hg_2^{2+} + 4AlCl_4^- \rightleftharpoons Hg_2Cl_2 + 2Al_2Cl_7^- . \qquad \text{(III-16)}$$

Donors of H^+ ions, which are acceptors of Cl^-, are also chloroacids:

$$H^+ + 2AlCl_4^- \rightarrow HCl + Al_2Cl_7^- . \qquad \text{(III-17)}$$

The extent of the pCl^- scale is limited towards basic media by the solubility of NaCl at the temperature used (pCl^- minimal) – which is defined by the point corresponding to that temperature on the liquidus of the phase diagram for $AlCl_3$ and NaCl mixtures. Towards acidic media, the highest values of pCl^- are reached for high concentrations of aluminium chloride in $NaAlCl_4$, and given by the expression:

$$pCl^- = pK_S + \log[Al_2Cl_7^-]. \qquad \text{(III-18)}$$

Experimental investigations have shown that, at 175 °C, pK_S=5.5 and pCl^- minimal=1.3; at 240 °C, pK_S=4.75 and pCl^- minimal=1.0 (concentrations expressed in mol kg^{-1}) (Trémillon and Létisse, 1968).

OXOACIDITY IN MOLTEN SALTS WITH AN OXO-ANION

In molten salts where the anion is an oxo-complex – like sulphates, carbonates, nitrates, phosphates, borates, silicates, etc. – , the concept of oxoacidity defined by Lux and Flood follows equally well as a consequence of the concept of solvoacidity.

1. In molten alkali sulphates, the reasoning is summarized as follows. The sulphate anion can dissociate, liberating an oxide ion, in accordance with the equilibrium (at high temperatures):

$$2SO_4^{2-} \rightleftharpoons S_2O_7^{2-} + O^{2-}. \qquad \text{(III-19)}$$

In this medium, therefore, an equilibrium is established between the solvent anions and their dissociation products: disulphate ions $S_2O_7^{2-}$ and oxide O^{2-}. To the extent that this dissociation remains extremely weak – this is the case up to the region of 1000 °C – we can assume that the activity of SO_4^{2-} remains practically equal to unity, and can write the mass action law as applied to the above equilibrium in the form of an activity product:

$$K_S = [S_2O_7^{2-}]\,[O^{2-}]. \tag{III-20}$$

In the eutectic $Na_2SO_4 + K_2SO_4$ at 950 °C, the value of pK_S is 9.4 (mole fraction scale) (Lux, 1948–1949). Thus, the activity of $S_2O_7^{2-}$ (or its concentration) is always connected, in this medium, with that of oxide ions O^{2-}. If we add O^{2-}, $S_2O_7^{2-}$ must disappear, and its concentration fall; if we add disulphate, on the other hand, it is the O^{2-} concentration which must fall; pO^{2-} can thus vary, from values close to zero obtained in a concentrated solution of alkali oxide (strongly basic medium), up to the highest values, close to pK_S, obtained in a concentrated solution of disulphate (strongly acidic medium). It is the oxide-donor nature of the sulphate ion which causes the limitation of the oxoacidity scale – to an extent of about 9 units, under the experimental conditions of Lux's work – and makes $S_2O_7^{2-}$ the strongest oxoacid existing in the medium.

The acidic and basic ranges are separated by a point of neutrality corresponding to $pO^{2-} = \frac{1}{2}pK_S = 4.7$.

In connection with the acido-basic properties of solutes in molten sulphates, it has been shown for example that metallic cations which are ready acceptors of O^{2-} ions, like Al^{3+}, Bi^{3+}, Zr^{4+} (introduced in the form of chlorides), behave as strong oxoacids, the solution of which leads to quantitative formation of the corresponding oxides (insoluble) and of $S_2O_7^{2-}$; but metaphosphate, PO_3^-, is a weak oxoacid, $pK_A=4.1$ – the corresponding oxobase being partially destroyed, neutralized by $S_2O_7^{2-}$.

2. In molten alkali carbonates, it is the CO_3^{2-} anion which intervenes by its dissociation, liberating carbon dioxide and again an oxide ion in accordance with the system:

$$CO_3^{2-} \rightleftharpoons CO_2 + O^{2-}. \qquad \text{(III-21)}$$

A similar argument to that applied to sulphates leads us to consider that CO_2 is the strongest oxoacid in molten carbonates and that the pO^{2-} scale is limited to the values obtained in presence of carbon dioxide under the highest partial pressures P_{CO_2}. In fact, the above equilibrium constant can be put in the form:

$$K_S' = P_{CO_2} \cdot [O^{2-}], \qquad \text{(III-22)}$$

whence

$$pO^{2-} = pK_S' + \log P_{CO_2}. \qquad \text{(III-23)}$$

pO^{2-} can thus be fixed by control of the partial pressure of carbon dioxide in presence of the molten carbonate.

Acceptors of O^{2-} which are stronger than CO_2 will behave as strong oxoacids in molten carbonates; this is the case for SO_3, PO_3^-, SiO_2 and numerous metallic ions. Some amphoteric metallic oxides can also dissolve as a consequence of their oxoacid character (Flood *et al.*, 1947).

3. In the case of molten alkali nitrates, a process similar to dissociation of NO_3^- ions has been suggested, leading to the nitronium cation, although of only temporary existence, being regarded as the strongest oxoacid existing in this solvent (Duke and Iverson, 1958–1959).

The case of molten alkali phosphates, silicates and borates can be treated in a quite similar manner, acidity being given the same definition (oxoacidity) and the same measure by the quantity pO^{2-}; but the most acidic media are different because of the limitation of the pO^{2-} scale by a different solvent-system, with a different constant of equilibrium – apart from the effect of differences in the utilization temperature.

Note. An evaluation of the accessible range of acidity in certain of these liquid media can be made by means of thermodynamic data, the standard Gibbs energy change of reactions.

For example, the standard Gibbs energy of formation of liquid sodium carbonate ΔG^*, starting from solid sodium oxide and carbon dioxide, enables us to calculate a value for the constant:

$$pK_S^* = \frac{-\Delta G^*}{2.3RT}. \qquad \text{(III-24)}$$

It is easily shown that this value corresponds to the number of pO^{2-} units existing, at the temperature under consideration, between the solution saturated of sodium oxide – the most basic solution possible – and the (acidic) solution saturated of carbon dioxide under the pressure 1 atm.

For the molten ternary eutectic $LiCO_3+NaCO_3+$ $+K_2CO_3$, Ingram and Janz (1965) have calculated the value $pK_S^* \approx 11$ at 600 °C. In the same manner, Rahmel (1968) has calculated a value $pK_S^*=19.7$ for the ternary eutectic $Li_2SO_4+Na_2SO_4+K_2SO_4$ also at 600 °C; this represents the interval of pO^{2-} between the solution saturated of oxide and the solution (very acidic) in equilibrium with sulphur trioxide under the pressure 1 atm ($S_2O_7^{2-}$ being itself in equilibrium with SO_4^{2-} and gaseous SO_3). Bombara *et al.* (1968) give a value very close to 20 for the eutectic of Li_2SO_4 and K_2SO_4 at the same temperature. Finally, Eluard (1970) has calculated the constant $pK_S^*=13.6$ for the eutectic NaOH+KOH at 227 °C, by means of the standard Gibbs energy of formation of liquid alkali hydroxides starting from solid alkali oxides and water vapour; this represents the interval of pH_2O, or of pO^{2-} (the two quantities can be defined here) between the saturated solution of oxide and the solution in equilibrium with water vapour under the pressure 1 atm.

In order to turn the value of pK_S^* thus calculated into that of the product of the concentrations of the oxide and strong oxoacid conjugate to the solvent-anion, close to the product of activities K_S, it is sufficient in principle to know, on one hand the relation between the concentration of the strong oxoacid (coefficient of Henry's law, if this is obeyed), and on the other hand the solubility of the (alkali) oxide at the temperature under consideration. For example, in the case of the NaOH+KOH eutectic at 227 °C, since experiment gives the value of 0.1 M for the solubility of the oxide and the relation $P_{H_2O} \approx 10^{-3}$ $[H_2O]$, it can be calculated:

$$pK_S = pK_S^* - 3 + 1 = 11.6. \qquad \text{(III-25)}$$

This value is identical with the value determined experimentally (cf. Section II.4, Equation II-68).

SOLVOACIDITY IN ACETIC ANHYDRIDE AND ACETYL CHLORIDE

As a last example of the use of the solvoacidity concept, which may concern reactions in organic chemistry, let us consider two aprotic molecular solvents with a certain relationship between them: acetic anhydride (ε=22 at 20 °C) and acetyl chloride (ε=15 at 20 °C).

So as to explain certain of the ionic exchange reactions in these solvents, the existence of self-ionization equilibria has been assumed; in acetyl chloride:

$$AcCl \rightleftharpoons Ac^+ + Cl^- \qquad \text{(III-26)}$$

(the symbol Ac^+ represents the acetyl cation CH_3CO^+); in acetic anhydride:

$$Ac_2O \rightleftharpoons Ac^+ + AcO^-. \qquad \text{(III-27)}$$

In this latter case, the activity product:

$$K_S = [Ac^+][AcO^-]$$

would be of the order of magnitude 3×10^{-15} (Jander *et al.*, 1948).

These solvent-systems make it possible to define solvoacids and solvobases appropriate to these media. Since the characteristic cation of the solvent is Ac^+, solvoacids are the donors of this ion – acetyl perchlorate, fluoride, chloride, bromide, thiocyanate, sulphide and fluoroborate – together with the substances which tend to produce it by virtue of their power of accepting the anion of the solvent, that is: acceptors of chloride in AcCl, acceptors of acetate AcO^- in Ac_2O – this is the case for metallic ions, $AlCl_3$, BF_3, $FeCl_3$, $SbCl_5$, SO_3 and naturally the hydrogen ion. Solvobases are the donors of Cl^- in AcCl, of AcO^- in Ac_2O (since these anions are the strongest bases which can exist there) – chloride or acetate complexes of metallic ions are the principal solvobases – together with substances which release these anions by virtue of their power of accepting the acetyl cation – pyridine, quinoline, etc.

Various neutralization reactions between solvoacids and solvobases have been described, which for strong solvoacids and strong solvobases amount to:

$$Ac^+ + Cl^- \rightarrow AcCl \quad \text{in acetyl chloride} \qquad \text{(III-28)}$$

$$Ac^+ + AcO^- \rightarrow Ac_2O \quad \text{in acetic anhydride.} \qquad \text{(III-29)}$$

Solvolytic reactions. Solvolytic reactions comparable to those produced in amphiprotic solvents (cf. Section III.2) – corresponding to exchanges between ions of the solvent and ions of the solute – are observed in ionizable aprotic molecular solvents such as Ac_2O and AcCl.

In acetyl chloride, for example, complex mercuric acetate undergoes a solvolytic decomposition, leading to formation of mercuric chloride and acetic anhydride; in the same manner, metallic oxides undergo a solvolysis, leading to metallic chlorides (complexes) and to acetate ions (which can form mixed chloro-acetate complexes).

One of the most characteristic solvolysis reactions is that in which water takes part (exchange H^+/Ac^+ and OH^-/Cl^- or AcO^-):

$$H_2O + AcCl \rightarrow AcOH + HCl \qquad \text{(III-30)}$$

$$H_2O + Ac_2O \rightarrow 2AcOH. \qquad \text{(III-31)}$$

Also, in acetic anhydride:

$$HX + Ac_2O \rightarrow AcX + AcOH. \qquad \text{(III-32)}$$

Various protonated compounds are capable of undergoing solvolysis reactions which can be interpreted in accordance with the same process; for example:

$$ROH + AcCl \rightarrow AcOR + HCl \qquad \text{(III-33)}$$

$$NH_3 + Ac_2O \rightarrow AcNH_2 + AcOH \qquad \text{(III-34)}$$

$$AcNH_2 + Ac_2O \rightarrow Ac_2NH + AcOH. \qquad \text{(III-35)}$$

CHAPTER IV

OXIDATION-REDUCTION PHENOMENA

In the same way that there are reactions in which ionic 'particles' such as H^+, Cl^- or O^{2-} take part, there are solution reactions in which electrons take part, and which are called 'oxidation-reduction reactions'. The theory of these reactions is very similar to that of ionic reactions, in particular that of Brønsted–Lowry; there is a parallelism between the definitions and between the methods of prediction.

REMINDER OF DEFINITIONS CONCERNING OXIDATION-REDUCTION PHENOMENA AND THEIR TREATMENT

A species which can give up one or more electrons is a reducing agent; on the other hand, a species which can absorb one or more electrons is an oxidizing agent. To every reducing agent there corresponds an oxidized form, and to every oxidizing agent there corresponds a reduced form; thus, we define oxidizing-reducing couples, which are acceptor-donors of electrons (e^-), and which can be symbolized by a relation of type:

$$\underset{\text{(acceptor)}}{\mathrm{Ox}} + ne^- = \underset{\text{(donor)}}{\mathrm{R}} . \tag{IV-1}$$

Oxidation and reduction reactions. Oxidation-reduction equilibria

The transformation which consists of removing electrons from a substance is an oxidation reaction; in the opposite direction, the transformation resulting in gaining of electrons is a reduction reaction.

These transformations can be brought about by exchange of electrons between a donor, reducing agent R_1 of a couple $Ox_1 + n_1 e^- = R_1$, and an acceptor, oxidizing agent Ox_2 of a couple $Ox_2 + n_2 e^- = R_2$. The exchange continues in principle until an oxidation-reduction equilibrium is reached, represented by:

$$n_2 R_1 + n_1 Ox_2 \rightleftharpoons n_2 Ox_1 + n_1 R_2 . \tag{IV-2}$$

For example*:

$$Fe^{II} + Ce^{IV} \rightleftharpoons Fe^{III} + Ce^{III}$$
$$3Fe^{II} + Cr^{VI} \rightleftharpoons 3Fe^{III} + Cr^{III}$$

* The Roman figures indicate the oxidation number of the element, which changes on transformation by oxidation or by reduction. The true species so designated is not necessarily the simple solvated ion (e.g. Fe^{3+}, Ce^{4+}, As^{3+}, etc), but it can also be a complex compound such as $Fe(CN)_6^{3-}$ for Fe^{III}, $HCrO_4^-$ for Cr^{VI}, $HAsO_3^{2-}$ for As^{III}, etc., in which it is known that the element affected by the transformation is uniquely that designated by its oxidation state. This means of expression holds good for quantitative treatment of transformation, on condition that all the other chemical factors – pH, concentrations of complexing reagents – remain constant.

$$3As^{III} + Br^{V} \rightleftharpoons 3As^{V} + Br^{-}.$$

By application of the mass action law, the equilibrium (IV-2) is characterized by an equilibrium constant K^0 which connects the activities of the four components. At a constant ionic strength, we can use an apparent constant K, as in the case of acid-base equilibria, connecting the concentrations of the substances concerned:

$$K = \left(\frac{[Ox_1]}{[R_1]}\right)^{1/n_1} \left(\frac{[R_2]}{[Ox_2]}\right)^{1/n_2} \qquad \text{(IV-3)}$$

Electrochemical reactions and equilibria

Transformations by oxidation or by reduction are still possible by means of an electrode, an electronic conductor dipped in the solution and able to exchange electrons with the oxidizing or reducing agents present (in solution) at its surface. There is electrochemical reduction when the electrode gives up electrons to an oxidizing agent: $Ox + ne^- \rightarrow R$; conversely there is electrochemical oxidation when a reducing agent gives up electrons to the electrode: $R - ne^- \rightarrow Ox$.

The constituents of an oxidizing-reducing system present at the surface of an electrode (the substance of which can possibly be one of these constituents: for example, systems of metal (electrode)-metallic cation) have a tendency to put themselves in equilibrium with it; the equilibrium concerned is electrochemical.

With two electrodes each one being the location of an

electrochemical equilibrium, we can produce a cell, of which we can measure the electromotive force E, that is the potential difference between the electrodes. The value of E is connected with the constant K of the oxidation-reduction equilibrium which corresponds to the two electrochemical equilibria; thus, the two electrochemical equilibria $Ox_1 + n_1 e^-$(electrode 1)$\rightleftharpoons R_1$ and $Ox_2 + n_2 e^-$ (electrode 2)$\rightleftharpoons R_2$, from which we can constitute a cell of electromotive force E, correspond to the oxidation-reduction equilibrium (IV-2), with the constant K, and we have:

$$E = E_2 - E_1 = \frac{2.3RT}{F} \log K + \frac{2.3RT}{n_2 F} \log \left(\frac{[Ox_2]}{[R_2]} \right)_{\text{electr. 2}} - \frac{2.3RT}{n_1 F} \log \left(\frac{[Ox_1]}{[R_1]} \right)_{\text{electr. 1}} \quad \text{(IV-4)}$$

(F is the Faraday, the quantity of electricity corresponding to one mole of electrons, about 96500 Coulombs, or alternatively 23.08 kcal $mol^{-1} V^{-1}$. At 20 °C, the coefficient $2.3RT/F$ has a value of 0.058 V).

When the activities of Ox_1 and R_1 at electrode 1, and of Ox_2 and R_2 at electrode 2, are equal to unity, the electromotive force has the 'standard' value:

$$E^0 = E_2^0 - E_1^0 = \frac{2.3RT}{F} \log K. \quad \text{(IV-5)}$$

Electrode potentials

To allocate values to the potential of an electrode where electrochemical equilibrium has been established requires agreement on a reference electrode to which we arbitrarily give the potential of zero. For aqueous solutions, the reference adopted is the normal hydrogen electrode, formed by a platinized platinum electrode where the electrochemical equilibrium $2H^+ + 2e^- \rightleftharpoons H_2$, is settled with activities equal to unity for H^+ (pH=0) and for H_2 (pressure $P_{H_2}=1$ atm). We then define the 'electrode potential' (with the symbol E) as the electromotive force of a cell of which electrode 2 is the electrode under consideration (system $Ox + ne^- \rightleftharpoons R$) and electrode 1 is the normal hydrogen electrode ($E_1=0$). By application of the relations (IV-2), (IV-3), (IV-4) and (IV-5), we then have:

$$E = E^0 = \frac{2.3RT}{nF} \log \frac{[Ox]}{[R]} \qquad \text{(IV-6)}$$

where:

$$E^0 = \frac{2.3RT}{F} \log K \qquad \text{(IV-7)}$$

K being the constant of the oxidation-reduction equilibrium:

$$\frac{1}{n} \mathrm{Ox} + \tfrac{1}{2}H_2 \rightleftharpoons \frac{1}{n} \mathrm{R} + H^+ \qquad \text{(IV-8)}$$

E^0 is called the 'standard' potential of the system $Ox + ne^- = R$.

Consequences for treatment of oxidation-reduction reactions

1. The above considerations show that each oxidizing-reducing system is characterized by a value E^0 of the standard potential. The different values of E^0 enable us to predict all possibilities of oxidation-reduction reactions, since the equilibrium constant (IV-2) is expressed, as a function of the standard potentials E_1^0 and E_2^0 of the two couples acting, by relation (IV-5). If $E_2^0 \gg E_1^0$, the reaction corresponding to (IV-2) in the direction from left to right is produced spontaneously by presence of R_1 and of Ox_2.

The classification order of E^0 values, as the values increase, corresponds to the order of increasing strength of oxidizing agents and to the order of decreasing strength of reducing agents. Thus, in aqueous solution ($+H_2SO_4$ 1M), Ce^{IV}, $E^0=1.44$ V, is a stronger oxidizing agent that Fe^{III}, $E^0=0.61$ V; Ce^{IV} therefore oxidizes Fe^{II} into Fe^{III}.

In the same way, Cr^{II}(Cr^{III}/Cr^{II}, $E^0=-0.37$ V) is a stronger reducing agent than tin (Sn^{II}/Sn, $E^0=-0.20$ V) and Cr^{II} reduces Sn^{II}.

2. A solution containing oxidizing and reducing agents is characterized by a value of the equilibrium potential equal to the electrode potential E. It is therefore expressed for the couple $Ox+ne^-=R$ by relation (IV-6); its value increases with the ratio of concentrations [Ox]/[R] on one

hand and with the value of E^0 of the system on the other hand.*

A medium with very high potential corresponds to an oxidizing medium. On the other hand, a medium with very low potential is a reducing medium.

The potential in a solution varies during the course of an oxidation-reduction reaction. Formula (IV-6) (or IV-9) enables us to calculate this variation, which characterizes quantitatively the reaction which has taken place. Experimental measurement of E, by means of an indicator electrode, enables it to be followed.**

* In a more general fashion, an oxidation-reduction reaction can call into play an oxidizing-reducing system represented by: $aA+bB+\ldots+ne^-=pP+qQ+\ldots$. For such a system, in which the different activities of the participants A, B,..., P, Q..., intervene, the equilibrium potential is expressed by the general formula (at constant ionic strength):

$$E=E^\circ+\frac{2.3RT}{nF}\log\frac{[A]^a\,[B]^b\ldots}{[P]^p\,[Q]^q\ldots}. \qquad \text{(IV-9)}$$

If certain constituents are saturated in the solution, we can give their activities the value unity. We can also include in the value of E^0 – which then becomes an apparent standard potential – the values of the activities, maintained constant, of certain participants.

** The measurement necessitates the use of a second reference electrode, that is one of known potential. There is in general a 'liquid junction' between the solution investigated and that which takes part in the system of the reference electrode; then, there is produced a 'junction potential' E_j which is included in the measured electromotive force. Provided that the same solvent is used for the two electrodes, this junction potential can be rendered negligible or approximately calculable.

Note. We can stress the analogy between the potential scale, which provides a measure of the oxidation-reduction phenomena, and the pH scale (or the scales of pO^{2-}, pCl^-, etc.), which provides a measure of the acid-base phenomena in a given solvent. It has even been suggested, to reinforce this analogy, that the potential E, expressed in Volts, should be replaced by the dimensionless quantity $pe^- = -(F/2.3RT)\,E$.

Standard potentials E^0 play the same part in prediction of oxidation-reduction reactions as do acidity constants pK_A of acid-base couples for prediction of reactions between acids and bases.

We must however note one essential difference, at least in the case of aqueous solutions and for numerous other solvents: unlike the proton, the electron is not a species which exists in the free solvated state, and its concentration cannot be defined. In addition, oxidizing and reducing agents dissolved in contact with the solvent are most frequently not in oxidation-reduction equilibrium with this latter.

This prevents the consideration of the existence, as in the case of acids and bases, of a solvolysis of oxidizing and reducing agents in solution, which would enable us to define a non-arbitrary origin for the scale pe^-.

Nevertheless, we can do this in the case of solvents which give rise to oxidation-reduction equilibria with their solutes, these solvents being essentially molten salts at high temperatures.

IV.1. Solvent Intervention in Oxidation-Reduction Phenomena

The above reasoning can be applied in principle to any solvent.* In order to establish a scale of potential in a given solvent, it is necessary to take the following steps:

1. Selection of a reference electrode the potential of which is taken arbitrarily as the origin of the scale. By analogy with aqueous media, the choice of the normal hydrogen electrode in the solvent under consideration (pH$(S)=0$, the other characteristics remaining unchanged: $P_{H_2}=$ $=1$ atm, platinized platinum electrode) is to be preferred when possible: molecular solvents such as alcohols, amides, carboxylic acids, DMSO, HMPA, liquid ammonia, hydrazine, etc.

For solvents in which the hydrogen electrode is not possible, another reference electrode must be selected. Since this choice is arbitrary, any electrode system giving an electrochemical equilibrium can in principle be used. For example, we shall see in the following chapter how the ferricinium$^{+}+e^{-}$ (platinum)$\rightleftharpoons$ferrocene is used, the standard potential of which can be adopted as the origin value of the potential scale in any solvent where this system operates. In molten alkali salts, we can use the alkali metal-alkali cation (see Section IV.2), etc.

* In molecular solvents of low dielectric constant, it is necessary to complete it so as to allow for ionic associations (ion-pairs), as in the case previously mentioned of acid-base properties (Section II.3).

2. Determination, with regard to the selected reference electrode, of the standard potentials of all the oxidizing-reducing systems existing in the solvent under consideration.

There are notable differences in oxidizing-reducing properties between one solvent and another. In the first place, identical oxidizing-reducing couples exhibit differing changes in standard potentials; Tables VII and VIII show some values which illustrate this fact. In the second place, the oxidizing-reducing systems which can be observed can differ from one solution to another, due to the fact that oxidation states of elements stable in one solvent can disappear in another. For example, the Cu^+ ion, which is not stable in aqueous solution, exists in solution in ethylenediamine, in molten dimethylsulphone, or in molten alkali chlorides or hydroxides; on the contrary, the Hg_2^{2+} ion which exists in aqueous solution, does not exist in ethylenediamine. In presence of an acidicaqueous solution, the stable oxidation states of manganese are Mn^{II}, Mn^{IV} and Mn^{VII}; in molten hydroxides these are Mn^{II}, Mn^{III} and Mn^{V}, etc.

The possibilities of oxidation-reduction reactions are therefore sometimes very different from one solvent to another.

Limitations of the potential scale by oxidation and reduction of the solvent

1. Let us consider aqueous solutions in the first place.

TABLE VII

Standard potential values (in Volts) of some oxidizing-reducing couples in different molecular solvents.
(Reference: normal hydrogen electrode in the solvent.
Concentrations in mol kg^{-1}.
Temperature: 25 °C unless stated to the contrary).

Solvent / Couple	Water	Meth-anol	Aceto-nitrile	Formic acid	Form-amide	Ammonia (at—35 °C)
Cs^+/Cs	—2.92		—3.16	—3.44		—1.95
Rb^+/Rb	—2.92	—2.91	—3.17	—3.45	—2.85	—1.93
K^+/K	—2.92	—2.92	—3.16	—3.36	—2.87	—1.98
Na^+/Na	—2.71	—2.73	—2.87	—3.42		—1.85
Li^+/Li	—3.05	—3.10	—3.23	—3.48		—2.24
Ca^{2+}/Ca	—2.87		—2.75	—3.20		—1.74
Zn^{2+}/Zn	—0.76	—0.74	—0.74	—1.05	—0.76	—0.53
Cd^{2+}/Cd	—0.40	—0.43	—0.47	—0.75	—0.41	—0.20
Tl^+/Tl	—0.34	—0.38			—0.34	
H^+/H_2	0	0	0	0	0	0
$AgCl/Ag$	0.22	—0.01			0.20	
Cu^{2+}/Cu	0.34		(—0.28)	—0.14	0.28	0.43
Cu^+/Cu	(0.52)	0.49	—0.38			0.41
Hg_2^{2+}/Hg	0.80			0.18		
Ag^+/Ag	0.80	0.76	0.23	0.17		0.83
Ferricinium$^+$/ ferrocene	0.40	0.41	0.19		0.54	
Cobalticinium$^+$/ cobaltocene	—0.92	—0.92	—1.15		—0.79	

According to Strehlow, 1966.

TABLE VIII

Standard potential values (in Volts) of some oxidizing-reducing couples in molten halides.
(Common arbitrary reference: Ag^+/Ag system. Mole fraction scale)

Molten halide	Ethylammonium chloride at 127°C	Ethylpyridinium bromide at 127°C	LiCl + KCl eutectic at 450°C	NaCl + KCl (1 : 1) at 700°C	NaBr + KBr (1 : 1) at 700°C	NaI at 700°C	NaF + KF eutectic at 850°C
Li^+/Li			-2.77				
Na^+/Na				-2.64	-2.27	-1.93	
Mg^{2+}/Mg			-1.94				
Al^{3+}/Al			-1.16		-0.55	-0.26	-2.14
Zn^{2+}/Zn			-0.93	-0.86	-0.68	-0.45	
V^{2+}/V			-0.89				
Cr^{2+}/Cr			-0.79	-0.76			
Cd^{2+}/Cd			-0.68	-0.62	-0.48	-0.31	
Fe^{2+}/Fe			-0.53	-0.52	-0.46	-0.07	
Pb^{2+}/Pb			-0.46	-0.39	-0.22	-0.07	
Sn^{2+}/Sn	-0.54	-0.29	-0.44	-0.37	-0.12	-0.11	
Co^{2+}/Co			-0.35	-0.32	-0.18	0.01	-0.71
Cu^+/Cu	-0.24	-0.16	-0.21	-0.26	0.08	0.04	-0.16
V^{3+}/V^{2+}	-0.47		-0.21				
Ni^{2+}/Ni			-0.16	-0.14	0.04	0.25	-0.64
Ag^+/Ag	0	0	0	0	0	0	0
Cr^{3+}/Cr^{2+}			0.01				
Bi^{3+}/Bi	-0.21	-0.02	0.05		0.31	0.25	
I_2/I^-			0.12			0.49	
Hg^{2+}/Hg	0.01	0.03	0.14		0.16	0.29	
Fe^{3+}/Fe^{2+}	0.33	0.62	0.62				
Cu^{2+}/Cu^+	0.43		0.69	0.60			
Br_2/Br^-			0.78		0.71		
Au^+/Au	0.73	0.67	0.95				
Cl_2/Cl^-			0.86	0.84			

According to Charlot and Trémillon, 1963; Picard and Vedel, 1969–70, for molten ethylammonium chloride and ethylpyridinium bromide.

When a solution of a non-electroactive electrolyte (that is one which is neither oxidizable nor reducible) is subjected to the action of an electrode which is brought to a decreasing potential (with respect to a reference electrode), starting from a certain potential a reduction of the water molecules is produced at the surface of the electrode, with formation of hydrogen and hydroxide ions. The oxidizing-reducing system is the following:

$$2H_2O + 2e^- = H_2 + 2OH^-, \qquad E^0 = -0.81\ \text{V}. \qquad \text{(IV-10)}$$

The existence of this system means that we cannot have a potential below a limit which corresponds to destruction of the solvent by reduction. The strongest reducing agents, such as alkali and alkaline earth metals, cannot exist in water, which they reduce with liberation of hydrogen and formation of hydroxide. In consequence of this, the conjugate oxidizing agents (alkali cations) are irreducible. Water therefore brings a limitation on the possibilities of existence of the strongest reducing agents and on the reducibility of the weakest oxidizing agents.

In other respects, if the solution is subjected to the action of an electrode (which cannot be attacked) brought to a progressively increasing potential, starting from a certain potential oxidization of the water molecules takes place, with formation of oxygen and of hydrogen ions. The system is the following:

$$2H_2O - 4e^- = O_2 + 4H^+, \qquad E^0 = 1.23\ \text{V}. \qquad \text{(IV-11)}$$

The existence of this system means that we cannot have potential greater than a limit which corresponds to destruction of the solvent by oxidation.

The strongest oxidizing agents, such as fluorine, cannot exist in water, which they oxidize with release of oxygen and formation of acid. In consequence of this, the conjugate reducing agents (F^- ions) cannot be oxidized. Therefore water brings a limitation to the possibilities of existence of the strongest oxidizing agents and to the extent of oxidization of the weakest reducing agents.

The range of potential corresponding to oxidation-reduction reactions which can be carried out in water is therefore restricted within these two limits. Its extent is given theoretically by the electromotive force of the cell corresponding to the oxidation-reduction equilibrium:

$$H_2 + \tfrac{1}{2}O_2 \rightleftharpoons H_2O\,, \qquad E^0 = 1.23\ \text{V}\,. \qquad \text{(IV-12)}$$

This is also called the 'decomposition potential' of water. This equilibrium quantity, which defines the theoretical extent of the scale of potential in water, plays the same part as the autoprotolysis constant, which defines the extent of the pH scale.

In practice, due to the slow rates of the two systems (IV-10) and (IV-11), and the absence of oxidation-reduction equilibria with water, the actual extent of the potential range in aqueous solution is greater than the theoretical value, being about 2 V (35 units of pe^-).

2. This double limitation on the scale of potential, on one

hand by the reduction system of the solvent, and on the other hand by its system of oxidation, is found also in other solvents. Thus it delimits a usable range of potential, the extent of which varies with the nature of the solvent.

The systems of reduction and oxidation of the various solvents are indeed very different from each other. However, some of them show similarities. In particular, amphiprotic solvents provide a reduction system leading, like that of water, to formation of hydrogen and the strong conjugate anion base of the solvent:

$$2HS + 2e^- = H_2 + 2S^- . \qquad \text{(IV-13)}$$

For example:

$$2CH_3CO_2H + 2e^- = H_2 + 2CH_3CO_2^-$$
$$2ROH + 2e^- = H_2 + 2RO^-$$
$$2HF + 2e^- = H_2 + 2F^- .$$

We therefore deduce qualitatively that as HS becomes an intrinsically weaker acid, the molecule of HS becomes more difficult to reduce and the solvent therefore enables us to attain more reducing potentials.

In the same way, the more difficult it is to oxidize the solvent the higher the potential it enables us to reach. Thus, acetonitrile enables the ClO_4^- anion to be oxidized (into the radical $ClO_4^{\bullet}$), which does not take place in water.

Other similarities between oxidation or reduction systems can be seen, enabling qualitative comparisons to be made. Thus, carboxylic acids give on oxidation the reaction known as Kolbe's (decarboxylation and formation of hydrocarbons and carbon dioxide).

In molten salts, the limitations are due to reduction or to oxidation either of the anion or of the cation. For example, in molten alkali halides, the limiting factor in the direction of low potentials is reduction of the metallic cation, but in molten alkali nitrates, the reduction of the NO_3^- anion (to NO_2^-) intervenes, since this reduction is easier than that of alkaline cations.

It is almost entirely restricted to molten salts at high temperatures, for the practical scale of accessible potential to correspond to the theoretical value of the decomposition voltage (the solvent follows the law of equilibrium). As in the case of water, molecular solvents at ordinary temperatures do not give rise to oxidation-reduction equilibria; their ranges of potential are determined by experiment.

Note. We saw in the first chapter that certain solvents have the property of solvating the electron. The strongest reducing agents, the alkali metals (M), are soluble in them, and dissolve into M^+ cations and solvated electrons e^-. When these solvents (ammonia, amines, HMPA, etc.) are subjected to the action of an electrode brought to progressively lower potentials, the limitation which appears is due, not to reduction of the solvent molecule, but to formation of dissolved free electrons. These are intrinsically the lowest potentials that can be reached.

These solvated electrons are not in fact in equilibrium

with the solvent, and disappear in time (more rapidly in in presence of a catalyst) by reduction of the solvent molecule; for example, for liquid ammonia:

$$NH_3 + e^- \rightarrow NH_2^- + \tfrac{1}{2}H_2 . \qquad \text{(IV-14)}$$

We can also consider that reduction of alkali ions in their molten halides, to the metal which partly dissolves, corresponds also to formation of solvated electrons in equilibrium with the molten salt (which enables us in this special case to put $pe^- = -\log[e^-]$).

IV.2. Influence of Acidity on Oxidation-Reduction Phenomena

Various ionic species are involved with electrons in most actual oxidizing-reducing systems, when transformation by oxidation or reduction brings with it a modification in composition of the chemical groups, either reducing or oxidizing. For example, the couples Fe^{3+}/Fe^{2+}, $Fe(CN)_6^{3-}/Fe(CN)_6^{4-}$, ferricinium$^+$/ferrocene, MnO_4^-/MnO_4^{2-}, Ag^+/Ag, etc., only set electrons in action, but the same cannot be true for the couples MnO_4^-/MnO_2, MnO_2/Mn^{2+}, $Ag(CN)_2^-/Ag$, $HgCl_4^{2-}/Hg_2Cl_2$, benzoquinone/hydroquinone, etc, which necessitate simultaneous use of H^+, OH^-, O^{2-}, CN^-, or Cl^- ions. It is self-evident that the activities of these ionic species in solution have an effect on the course of the oxidation-reduction phenomena in which such systems are implied, and that changes in them bring about modifications to the apparent potential values which characterize these systems.

Acidity is the essential feature whose effect must be considered on oxidation-reduction phenomena. In the traditional case of acidity in the sense of the Brønsted-Lowry definition, it is there the variations in pH which are capable of modifying the equilibrium potentials. To demonstrate this effect, we use potential-pH equilibrium diagrams, on which are shown the changes in equilibrium potential of an oxidizing-reducing system as a function of pH, for given activities of the oxidizing agent and of the reducing agent (remembering that species in saturated solution have a constant activity, which we may take as equal to unity).

In the case of a solvent leading to a non-protonic definition of acidity, we shall have recourse in the same way to equilibrium diagrams of potential against acidity, that is potential-pCl^- if dealing with chloroacidity, potential-pO^{2-} if oxoacidity, etc.

So as to show the trend of these diagrams, how they are obtained and applied, we will briefly describe some characteristic examples, first recalling the principle of potential-pH in aqueous solution.

Potential-pH equilibrium diagrams in aqueous solution and in similar solvents

1. The first point is to define the bounds of the diagram,

that is the range of potential and acidity inside which oxidation-reduction phenomena can take place in the solvent under consideration.

For aqueous solutions, it is known that this range is limited both by acidity considerations and by considerations on the oxidation-reduction potential. The limitations to this latter correspond to systems (IV-10) and (IV-11); it is easy to show that the characteristic equilibrium potentials of these systems vary with pH. We have in fact for (IV-10) (at 20 °C):

$$E/\mathrm{V} = -0.81 - 0.029 \log(P_{\mathrm{H_2}}[\mathrm{OH^-}]^2). \qquad \text{(IV-15)}$$

Having regard to the ionic product of water, let:

$$E/\mathrm{V} = -0.029 \log P_{\mathrm{H_2}} - 0.058\ \mathrm{pH}. \qquad \text{(IV-16)}$$

For a hydrogen pressure $P_{\mathrm{H_2}}=1$ atm (it is assumed that the solution is saturated with this gas under the pressure applied), $E=-0.058$ pH. The equilibrium potential of the system which limits the scale of potential therefore varies from 0.0 V at pH=0 (reduction $2\mathrm{H^+}+2e^- \rightarrow \mathrm{H_2}$) to -0.81 V at pH=14.

For system (IV-11) (at 20 °C):

$$E/\mathrm{V} = 1.23 + 0.015 \log(P_{\mathrm{O_2}}[\mathrm{H^+}]^4) =$$
$$= 1.23 + 0.015 \log P_{\mathrm{O_2}} - 0.058\ \mathrm{pH}. \qquad \text{(IV-17)}$$

For an oxygen pressure $P_{\mathrm{O_2}}=1$ atm, the equilibrium potential of the second system which limits the scale of potential varies therefore from 1.23 V at pH=0, to 0.42 V at pH=14.

In this manner, the area of E-pH which can be used in an aqueous solution is shaped like a parallelogram, with two sides parallel to the potential axis corresponding to the limitations on the pH scale (towards 0 and towards 14), and two parallel sides 1.23 V apart and varying as -0.058 pH, corresponding to the limitations on the potential scale.

The distance between these two latter limitations is, for the reasons already mentioned, slightly greater than the theoretical value of 1.23 V (about 2 V), if we are concerned with the region of potential and acidity accessible in practice.

In the same way, for other amphiprotic solvents, the accessible region of potential and acidity is defined from the extent of the scales of pH (pK_S units) and of potential, and from the variation in equilibrium potentials of the limiting systems of the solvent as functions of pH. The same law applies to the limit on the reducing side: variation from $E=0.0$ V at pH=0, to $E=-0.058$ pK_S at pH=pK_S (slope of graph: 0.058 V per unit of pH) (System IV-13). On the oxidation side, a theoretical (and practical) variation of the limit as a function of pH takes place in general in the same way.

2. The effect of pH on the equilibrium potential of an oxidizing-reducing system which can be observed in solution is described by a graph drawn within the E-pH region so defined.

A change in the equilibrium potential with pH is produced when this system brings H^+ ions into play. For example:

$$MnO_4^- + 3e^- + 4H^+ = MnO_2(s) + 2H_2O \quad \text{(IV-18)}$$

$$E/V = 1.68 + 0.020 \log [MnO_4^-] - 0.077 \text{ pH} \quad \text{(IV-19)}$$

$$\text{benzoquinone} + 2e^- + 2H^+ = \text{hydroquinone} \quad \text{(IV-20)}$$

$$E/V = 0.70 + 0.058 \log \frac{[\text{benzoquinone}]}{[\text{hydroquinone}]} - 0.058 \text{ pH}. \quad \text{(IV-21)}$$

Modifications to the law of variation as a function of pH (changes of slope) are produced when the oxidizing and/or reducing agent present has several acid-base forms according to the pH of the solution. For example, for the couple As^V/As^{III} we have:

$$E = E^{0\prime} + 0.058 \log \frac{[As^V]}{[As^{III}]}. \quad \text{(IV-22)}$$

$E^{0\prime}$ varies with pH in a different manner depending on whether the As^V is in the form of arsenic acid H_3AsO_4 ($pH \ll 2$) or in the form of one of its conjugate bases $H_2AsO_4^-$, $HAsO_4^{2-}$, AsO_4^{3-}, and whether the As^{III} is in the form of arsenious acid H_3AsO_3 ($pH \ll 8$) or in the form of one of its conjugate bases.

The influence of acidity can be very different from one

solvent to another – especially in the case where it is due to the formation of complexes between the oxidizing or the reducing agent and the S^- anions of an amphiprotic solvent (cf. acidity of metallic cations, Section II.2). Thus the equilibrium potential of metal-metallic ion couples varies with pH along with formation of hydroxide when in water, of acetate in acetic acid, of amide in ammonia, of acetamide complexes in molten acetamide, etc., which brings with it a very different behaviour of these couples depending on the solvent under consideration.

Potential-pH_2O equilibrium diagrams in molten alkali hydroxides

We have already seen (Section II.4) that acidity (in the traditional sense) is measured in molten alkali hydroxides by the quantity pH_2O. All oxidizing-reducing systems involving water molecules (or O^{2-} ions) will undergo variations in potential when pH_2O varies.

1. The region E-pH_2O delimited by the properties of the solvent is defined in the following manner. The lower limit of the potential scale is set, in an acidic medium, or in a neutral medium, by reduction of water (strong acid) to hydrogen. The equilibrium potential corresponding to this system is given by:

$$E = E^0_{H_2/H_2O} + \frac{2.3RT}{2F} \log \frac{[H_2O]^2}{P_{H_2}}$$

$$= E^0_{H_2/H_2O} - \frac{2.3RT}{2F} \log P_{H_2} - \frac{2.3RT}{F} pH_2O \qquad \text{(IV-23)}$$

which is the potential of a hydrogen electrode. This equilibrium potential is represented graphically by a series of straight lines corresponding to different values of P_{H_2}, with a variation depending on $-(2.3\,RT/F)\,pH_2O$ ($=-0.10\,pH_2O$ at the temperature of 227 °C; investigations in the eutectic NaOH+KOH, Goret, Eluard and Trémillon, 1967–70). Under conditions of the normal hydrogen electrode, $E = E^0_{H_2/H_2O}$.

At about $pH_2O=8$, the potential of the hydrogen electrode becomes less than the potential, independent of pH_2O, which corresponds to reduction of the alkali cations Na^+ and K^+, which then provides, in basic medium ($pH_2O \approx 8$ to 10.5), the lower limit of potential. Alkali metals are appreciably soluble in their molten hydroxides. The arbitrary origin of potential adopted for this medium is the potential of an electrode in a (hypothetical) 1 M solution of alkali metal. With respect to this reference, it has been found that $E^0_{H_2/H_2O}=0.82$ V.

The upper limit of the potential scale is due to oxidation of hydroxide ions (or of free oxide ions in a basic medium). At the highest values of pH_2O, the limiting system is the oxide-peroxide couple, $2O^{2-} - 2e^- = O_2^{2-}$, to which there corresponds the equilibrium potential:

$$E/V = 2.55 + 0.05 \log [O_2^{2-}] - 0.10\,pH_2O \qquad \text{(IV-24)}$$

(all concentrations are expressed in mol l^{-1}).

In neutral or acidic media, peroxide ions undergo a disproportionation and the limiting system becomes the hydroxide-superoxide couple, $4OH^- - 3e^- = O_2^- + 2H_2O$, to which there corresponds the equilibrium potential:

$$E/V = 2.28 + 0.033 \log [O_2^-] - 0.066\, pH_2O \quad \text{(IV-25)}$$

(all concentrations are expressed in mol l^{-1}).

Overall, the range of E-pH_2O utilizable in the molten eutectic NaOH+KOH at 227 °C has the shape shown on the diagram of Figure 7.

2. The graphs of variation of equilibrium potentials of the oxidizing-reducing systems observable in molten hydroxides are to be found inside the region so defined. We will consider only the case of nickel, as an example.

As far as the couple $Ni^{II} + 2e^- = Ni(s)$ is concerned, the potential is independent of pH_2O when Ni^{II} is in the form of solvated Ni^{2+} ions, that is in a very acidic medium (values of pH_2O near to zero); $E^0 = 1.45$ V. An increase in the value of pH_2O leads to precipitation of oxide NiO – when the solubility product of this latter substance is reached –, and we must then consider the system:

$$NiO(s) + 2e^- + H_2O = Ni(s) + 2OH^-$$

of which the equilibrium potential varies according to

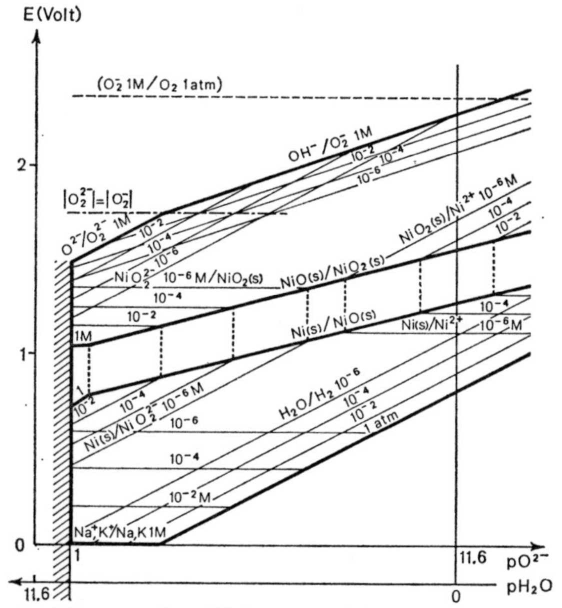

Fig. 7. Diagram of equilibrium potential against pH_2O (or pO^{2-}) in the molten eutectic NaOH + KOH at 227°C. (From Goret and Trémillon, 1968.)

the relationship:

$$E/V = 1.30 - 0.05\,pH_2O. \quad (IV\text{-}26)$$

Finally, in a very basic medium, NiO is redissolved in the state of NiO_2^{2-} anions and the oxidizing-reducing system is then:

$$NiO_2^{2-} + 2e^- + 2H_2O = Ni(s) + 4OH^-,$$

with:

$$E/V = 1.80 + 0.05 \log [NiO_2^{2-}] - 0.10\, pH_2O. \tag{IV-27}$$

This variation of equilibrium potential of the couple $Ni(s)/Ni^{II}$, for different values of the concentration of Ni^{II} when this latter is dissolved, is shown on the graph in Figure 7.

Experimental results also provide evidence for the existence of the couple $Ni^{IV}+2e^-=Ni^{II}$. Only one substance, the dioxide NiO_2, corresponds to Ni^{IV}; this dioxide is insoluble in the molten eutectic at 227 °C. The variations which can be calculated for the equilibrium potential of this couple as a function of pH_2O are shown in the same manner on the graphs in Figure 7, which therefore shows all the oxidizing-reducing and acid-base properties of nickel in the media considered.

POTENTIAL-pO^{2-} EQUILIBRIUM DIAGRAMS IN VARIOUS MOLTEN SALTS

In the previous chapter, we saw the interest of the oxoacidity concept in molten salts. The influence of oxoacidity on the oxidizing-reducing properties in these media at high temperatures is therefore of especial interest; this influence is shown in the equilibrium diagrams E-pO^{2-}, which exhibit the same features as E-pH (or E-pH_2O) previously considered.

The data for the graphs of E against pO^{2-} was obtained partly by means of experimental electrochemical measurements – in the molten eutectic LiCl+KCl, at 450 °C, for example (Delarue, 1960; Molina, 1961; Leroy, 1962) –, and partly with the assistance of classical thermodynamic data, the free enthalpies (Gibbs energies) of formation of various compounds starting from their constituents (at the temperature of the medium) – for example, for molten alkali chlorides (Littlewood and Edeleanu, 1960–1962), for molten alkali carbonates (Ingram and Janz, 1965) and for molten alkali sulphates (Rahmel, 1968), etc.; this method of obtaining data, in conjunction with experimental methods, has also been used for molten hydroxides (Eluard and Trémillon, 1968–1970).

1. Let us first consider the case of molten alkali chlorides (which we will denote by MCl, M being Li, Na, K, or a mixture of these elements in the case of molten eutectics).
(a) The useful range of E and pO^{2-} is defined as follows: pO^{2-} is in principle limited only in the direction of the most oxobasic media, by the solubility of the alkali oxide M_2O; we then have $pO^{2-}_{min} \approx -\log(\text{solubility of } M_2O)$.

The scale of potential is limited, on one hand by the system $M + e^- = M$ (pure) – the potential of which we will take as the origin (it is invariable once the molten salt is saturated with metal) – and on the other hand by the system $Cl^- = \frac{1}{2}Cl_2 + e^-$, the equilibrium potential of which is given by:

$$E = E^0_{\mathrm{Cl}} + \frac{2.3RT}{2F} \log P_{\mathrm{Cl}_2}. \tag{IV-27}$$

P_{Cl_2} represents the partial pressure of chlorine in equilibrium with the molten chloride; if the solutions in the molten salt remain sufficiently dilute, we assume that the activities of the constituent ions M^+ and Cl^- are in practice equal to unity. If ΔG_{MCl} is the standard free enthalpy of the reaction $\mathrm{M(pure)} + \frac{1}{2}\mathrm{Cl_2(gas)} \rightarrow \mathrm{MCl}$ (liquid), at the temperature under consideration, we calculate, from the relationship (IV-5) and the expression for the equilibrium constant K as a function of the standard free enthalpy of the reaction, that:

$$E^0_{\mathrm{Cl}} = -\frac{\Delta G_{\mathrm{MCl}}}{F}. \tag{IV-28}$$

The equilibrium potential of the molten solution in presence of chlorine, under different partial pressures P_{Cl_2}, is represented in the graph E-pO^{2-} by a series of straight lines parallel to the pO^{2-} axis (Figure 8). For molten NaCl at 800 °C for example, $\Delta G_{\mathrm{NaCl}} = -73.7$ kcal, whence $E^0_{\mathrm{Cl}} = 3.20$ V. The equilibrium potential under a pressure $P_{\mathrm{Cl}_2} = 10^{-6}$ atm is 2.55 V, whatever the value of pO^{2-}.

(b) The behaviour of oxygen – a very important element in connection with oxidation-reduction phenomena, especially at high temperatures – is represented as follows. Suppose that O_2 in molten alkali chlorides is reduced

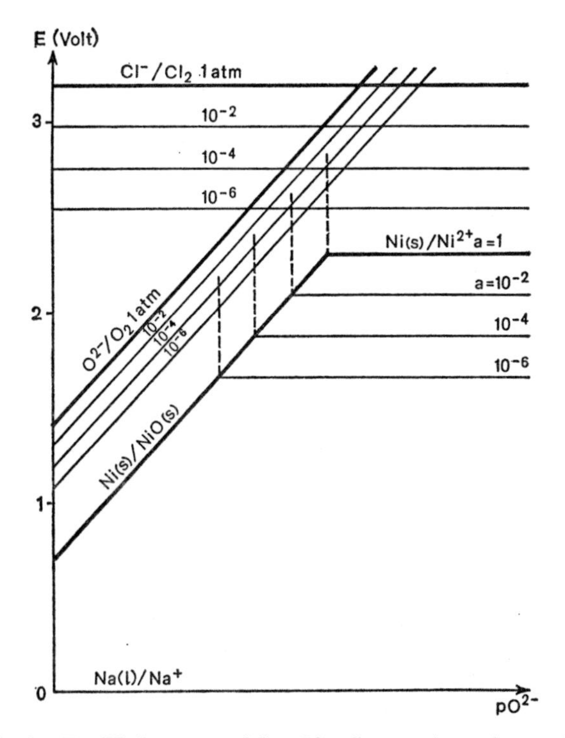

Fig. 8. Equilibrium potential- pO^{2-} diagram in molten sodium chloride at 800°C.

only in the state of the oxide ion O^{2-}. The equilibrium potential of a solution of M_2O (that is of free oxide ions O^{2-}) in presence of oxygen under partial pressure P_{O_2} is then given by the formula:

$$E = E^0_{O_2} + \frac{2.3RT}{4F} \log P_{O_2} + \frac{2.3RT}{2F}\, pO^{2-}. \qquad \text{(IV-29)}$$

For a given partial pressure P_{O_2}, this equilibrium potential increases linearly with pO^{2-}.

It is to be noted that Cl_2 oxidizes O^{2-} to oxygen in an oxobasic medium, whilst on the contrary it is oxygen which forms Cl_2 in a sufficiently oxoacidic medium. We can make use of this phenomenon to impose the oxoacidity of the medium; due to the oxidation-reduction equilibrium

$$\tfrac{1}{2}O_2 + 2Cl^- \rightleftharpoons \tfrac{1}{2}Cl_2 + O^{2-},$$

the application of partial pressures of chlorine and of oxygen 'buffers' the value of pO^{2-} (the activity of chloride remaining constant).

(c) The behaviour of a metal is represented as in the case of molten hydroxides. Let us consider again nickel as an example. The graph of variation of equilibrium potential of the system $Ni^{II}/Ni(s)$ is made up of two linear parts; one is independent of pO^{2-}, for the system Ni^{2+} (solvated by Cl^-)$+2e^- = Ni(s)$, for high values of pO^{2-}; the other, corresponding to the system $NiO(s)+2e^- = Ni(s)+$ $+O^{2-}$, for low values of pO^{2-}, expressing the relationship:

$$E = E^0_{Ni/NiO} + \frac{2.3RT}{2F}\, pO^{2-}. \qquad \text{(IV-30)}$$

The point of intersection between the two straight lines corresponds to the value of pO^{2-} at the commencement of precipitation of the oxide NiO (Figure 8).

We note that the lines showing variation of equilibrium potentials of the systems Ni/NiO and O_2/O^{2-} are parallel. The distance between them, equal to the difference of the standard potentials of the two systems, is also independent of the solvent, since it corresponds to the oxidation-reduction equilibrium:

$$\mathrm{Ni}(s) + \tfrac{1}{2}\mathrm{O}_2(g) \rightleftharpoons \mathrm{NiO}(s). \qquad \text{(IV-31)}$$

We can thus deduce this distance from the value of the standard free enthalpy ΔG_{NiO} of the formation reaction of the oxide NiO at the temperature under consideration:

$$E^0_{\mathrm{O}_2} - E^0_{\mathrm{Ni/NiO}} = -\frac{\Delta G_{\mathrm{NiO}}}{2F}. \qquad \text{(IV-32)}$$

The same holds for all metal/metallic oxide couples.

In the cases of many metallic elements, the phenomena are more complex because of the existence of several oxidation states. The E-pO^{2-} graph brings out the different variations of equilibrium potential between two neighbouring oxidation states. For example, Figure 9 shows the approximate experimental graphs obtained by Molina (1961) in the case of uranium and vanadium in the molten eutectic LiCl+KCl at 450 °C. Let us consider particularly the vanadium diagram. In addition to the metallic state, four positive oxidation states can be

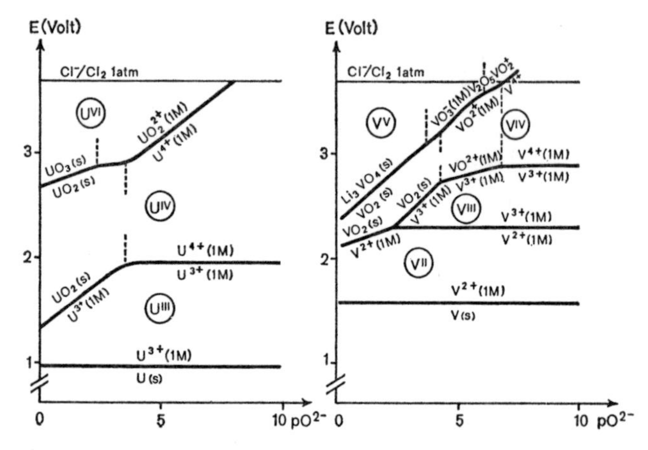

Fig. 9. Potential- pO^{2-} diagrams for uranium and vanadium in the molten eutectic LiCl+KCl at 450°C (from Molina, 1961).

found: V^{II}, V^{III}, V^{IV} and V^{V}. The cations V^{2+} and V^{3+} (solvated by Cl^-) do not appear to form an insoluble oxide, hence the potential of the couples V/V^{2+} and V^{2+}/V^{3+} remains independent of pO^{2-}. On the other hand, V^{4+} (added in the form of VCl_4, fairly volatile) can form successive oxocomplexes VO^{2+}, then VO_2 (insoluble) when the value of pO^{2-} is reduced, for example by addition of sodium oxide; V^{V} can exist only in the form of the successive oxocomplexes VO_2^+, V_2O_5, metavanadate VO_3^- and orthovanadate VO_4^{3-}, this latter forming the insoluble compound Li_3VO_4. In consequence the potential of the couple V^{IV}/V^{3+} starts to decrease with pO^{2-} in the most basic region where oxocomplexes of V^{4+} are formed; indeed the equilibrium potential corresponding to the system $VO^{2+}+e^- = V^{3+}+O^{2-}$ is:

$$E = E^0_{VO^{2+}/V^{3+}} + \frac{2.3RT}{F} \log \frac{[VO^{2+}]}{[V^{3+}]} + \frac{2.3RT}{F}\, pO^{2-} . \tag{IV-33}$$

The potential of the couple V^V/V^{IV} varies in the same direction as a function of pO^{2-}, since oxocomplexes of V^V are seen to be more stable than the homologous oxocomplexes of V^{IV} and, at the same value of pO^{2-}, V^V is found in an oxocomplex form which is richer in O^{2-} ions than V^{IV}. The equilibrium potential corresponding to the system $VO_2^+ + e^- = VO^{2+} + O^{2-}$, for example, is:

$$E = E^0_{VO_2^+/VO^{2+}} + \frac{2.3RT}{F} \log \frac{[VO_2^+]}{[VO^{2+}]} + \frac{2.3RT}{F}\, pO^{2-} . \tag{IV-34}$$

This influence of pO^{2-} on the equilibrium potentials of vanadium systems gives rise to a disproportionation of V^{3+} in a strongly basic medium (concentrated O^{2-}). On the other hand, the oxidizing power of V^V increases when pO^{2-} increases, to such a point that the compounds of vanadium(V) oxidize Cl^- to Cl_2 and cease to exist in presence of a sufficiently strong oxoacid, such as SO_3, PO_3^-, or Al^{3+}; conversely, it is possible to dissolve the dioxide VO_2 by oxidation (in the metavanadate state) by chlorine, in presence of a sufficiently strong oxobase such as carbonate ions.

2. In the case of molten alkali carbonates or sulphates, the essential difference lies in the delimitation of the accessible regions of E and pO^{2-}, since the graph showing the behaviour of the oxidizing-reducing systems is drawn as in the case of molten chlorides.

As for E-pH in water or for E-pH_2O in molten hydroxides, the accessible region in these media is limited both by considerations of oxoacidity and considerations of oxidation-reduction. The limitations on the scale of oxoacidity have been described in the previous chapter (Section III.3); it was shown that the values of the free enthalpy of the formation reactions of alkali carbonates or sulphates in the liquid state enable us to calculate the extent of the scale between the most basic media (saturated solution of M_2O) and the most acidic media (P_{CO_2} or $P_{SO_3}=1$ atm), being 11 units in the case of the ternary eutectic $Li_2CO_3+Na_2CO_3+K_2CO_3$ at 600 °C and 19.7 units in the case of the ternary eutectic $Li_2SO_4+Na_2SO_4+$ $+K_2SO_4$ at the same temperature.

On their part, the scales of potential are similarly limited in two ways. On the side of the high values, the limitation is caused by oxidation of the CO_3^{2-} or SO_4^{2-} anions; the system concerned is O^{-II}/O_2, either, in molten carbonates:

$$CO_3^{2-} = \tfrac{1}{2}O_2 + CO_2 + 2e^- \qquad \text{(IV-35)}$$

or, in molten sulphates:

$$2SO_4^{2-} = \tfrac{1}{2}O_2 + S_2O_7^{2-} + 2e^- . \qquad \text{(IV-36)}$$

In both cases, the equilibrium potential varies as a function of pO^{2-} as for the system O^{-II}/O_2 in molten chlorides, in accordance with the theoretical relationship (IV-29).

On the low value side, two limiting systems can be expected: either reduction of alkali cations to metal, system $M^+ + e^- = M$, of which the equilibrium potential is independent of pO^{2-}, or reduction of the anions CO_3^{2-} or SO_4^{2-}. In the case of carbonates, at a not very high temperature, the reduction is to graphite:

$$CO_3^{2-} + 4e^- = C(s) + 3O^{2-} \qquad \text{(IV-37)}$$

The equilibrium potential of this system varies with pO^{2-} according to the relationship:

$$E = E^0_{C/CO_3{}^{2-}} + \frac{3}{4} \cdot \frac{2.3RT}{F}\, pO^{2-} . \qquad \text{(IV-38)}$$

In the case of sulphates, reduction of SO_4^{2-} anions leads to formation of sulphide:

$$SO_4^{2-} + 8e^- = S^{2-} + 4O^{2-} . \qquad \text{(IV-39)}$$

The equilibrium potential of this system also varies with pO^{2-} in accordance with the relationship:

$$E = E^0_{S^{2-}/SO_4{}^{2-}} - \frac{2.3RT}{8F} \log\, [S^{2-}] + \frac{2.3RT}{2F}\, pO^{2-} . \qquad \text{(IV-40)}$$

By comparison of the standard free enthalpies of the formation reactions of liquid alkali carbonates, either starting from graphite, oxygen and alkali oxide, or from alkali metal, oxygen, and carbon dioxide, it appears that reduction to the graphite state must in theory take place more easily than deposition of the alkali metal; it is therefore the first factor which must limit the scale of potential, in accordance with relation (IV-38). By retaining, nevertheless, the (hypothetical) equilibrium potential of the couple M/M^+ as the arbitrary origin of potential, the calculations carried out by Ingram and Janz (1965), for the ternary eutectic at 600 °C, lead to values for the standard potentials of the two limiting systems:

$$E^0_{C/CO_3{}^{2-}} = 0.01\ \text{V} \quad \text{and} \quad E^0_{O_2} = 1.52\ \text{V}.$$

A similar calculation, carried out for molten sulphates by Rahmel (1968), shows that reduction of SO_4^{2-} ions to sulphide is also easier than deposition of the alkali metal, which leads to a limitation of the scale of potential in the direction of reducing media as given by relationship (IV-40). For the ternary eutectic at 600 °C:

$$E^0_{S^{2-}/SO_4{}^{2-}} = 2.01\ \text{V} \quad \text{and} \quad E^0_{O_2} = 2.88\ \text{V}$$

(with respect to the potential of the system Li/Li^+ taken as origin).

The utilizable regions appear as quadrilaterals on the graph E-pO^{2-} on Figure 10, which shows these limitations in the two media considered. If, in the case of molten

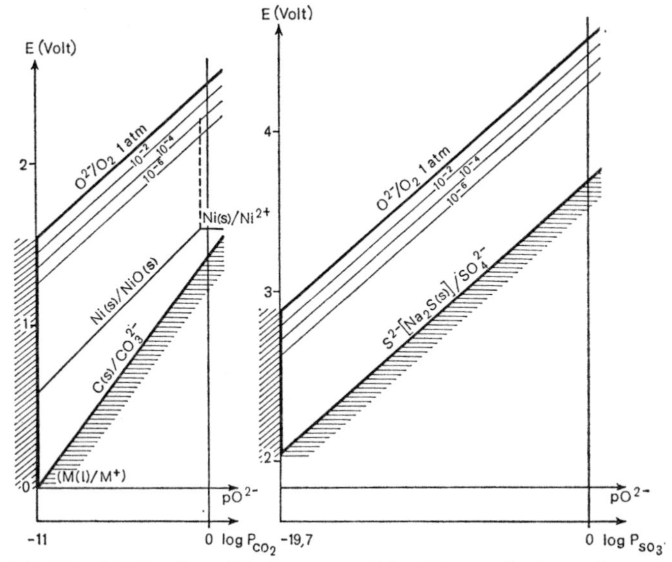

Fig. 10. Limit of equilibrium potential-pO^{2-} graphs in molten alkali carbonates and sulphates, at 600°C (from Ingram and Janz, 1965; Rahmel, 1968).

sulphates, the region is a parallelogram similar to the *E*-pH region in aqueous solution, the theoretical range of potential in molten carbonates appears more restricted in an oxoacidic medium than in an oxobasic medium, because the lower limit varies versus pO^{2-} with a slope greater than that of the upper limit.

Interest of potential-acidity equilibrium diagrams

Graphs of potential against acidity provide, in the first

place, an instantaneous overall view on the properties of an element in a given solvent: the different oxidation states which can exist and the influence of acidity on these states. The graphs of variation of equilibrium potentials set limits to the predominance zones of each oxidation state, inside which there also appear the different acid-base forms possible.

Examination and comparison of the diagrams for the various elements then shows the reactions possible between the constituents of several systems. Predictions concerning the conditions of production of these reactions follow from these. Diagrams showing only equilibrium values can naturally lead to information valid for equilibrium only. These throw no light on the kinetic aspect of the reactions. This is the main limitation of their interest, since in many cases the slowness of the reactions makes equilibrium magnitudes unattainable and also makes experimentally illusory the results deduced from the equilibrium data. In the case of molten salts however, the large temperature rise necessitated by the use of these particular solvents assists in establishment of equilibria, and the potential-acidity diagrams are then more reliable for predictions than are similar diagrams in aqueous solution or in molecular solvents at ordinary temperatures.

One of the chief applications of the potential-acidity graphs is concerned with prediction of conditions under which metals can be corroded, or rather, in practice, of conditions which ensure the protection of metals against corrosion by liquids such as water, organic solvents and molten salts and hydroxides. On this subject, it is useful to recall that Pourbaix (1949) suggested an arbitrary assumption that a metal is corroded if the concentration of ions in equilibrium becomes greater than 10^{-6}M; this hypothesis enables us to fix the region and hence the conditions of immunity of a metal, conditions that can be realized by addition of chemical reducing agents or by means of cathodic protection. In the case of non-aqueous media, the same hypothesis can of course be retained, with the same limits of validity.

Outside its zone of immunity, therefore, a metal runs the risk of being attacked. If the acidity of the medium is such that the metallic ions can exist in solution in appreciable concentration, there is no doubt as to the possibility of attack; but in the zone of acidity where an insoluble metallic compound is formed – especially an oxide, as we have seen in the case of molten salts and hydroxides –, the attack can be arrested if this oxide forms a coherent film over the surface of the metal, leading to passivation. The theoretical diagram does not enable us to distinguish whether the oxide formed leads to passivity and arrests corrosion or not; experimental results are necessary on this point. But the diagram can show the regions of existence of different oxides, corresponding to different oxidation states of the metallic element, and having possibly very different physical properties. If some of them lead to passivity, new metal protection zones will appear on the diagram and will determine the conditions suitable for obtaining this protective layer by means of chemical treatment or by electrochemical methods (anodization). In any circumstances, the determination of zones of formation of insoluble oxides is seen to be necessary before experimental investigations showing subsequently whether these oxides lead to passivity or not.

Let us consider by way of example the behaviour of nickel in contact with molten alkali hydroxides (at 227 °C). The immunity zone of the metal appears on the diagram of Figure 7. We see that complete immunity ceases to be possibile in a very acidic medium (very highly hydrated), where the nickel starts to be attacked by the water. Apart from water, the most important corrosive agent is of course atmospheric oxygen in contact with the metal and the molten medium. The oxygen clearly oxidizes the nickel, firstly to the state Ni^{II}, soluble in a medium either very acidic or basic, insoluble in a slightly acidic or neutral medium. But the final oxidation state is the dioxide NiO_2, which is insoluble whatever the acidity. Experiment shows that it forms an extremely thin, coherent film, rendering passive the surface of the metal; this film therefore protects nickel from an attack by oxygen and other strongly oxidizing agents (superoxide, peroxide) throughout almost all the acidity range (in the most basic medium, the metal might undergo a weak attack by concentrated peroxide, oxidation ceasing then when

NiO_2^{2-} is formed). This property of nickel explains its current use as a metallic container for molten alkali hydroxides, especially for oxidizing alkaline smelting.

In a similar manner, it has been shown that platinum, in contact with molten hydroxides, has a wide zone of immunity, but it is oxidized with formation of Pt^{2+} in a very acidic medium, of PtO_2^{2-} in a very basic medium, and of an insoluble oxide PtO, not leading to passivation, in the intermediate acidity range; as a result of this, platinum is rather strongly attacked. Gold is a metal which can be even more strongly attacked, especially in a basic medium, with formation of an aurate (I) ion AuO^{3-}, then an aurate (III) ion AuO_2^-, both being soluble; in an acidic medium, gold is oxidized directly to Au_2O_3, not leading to passivity, and having a solubility which becomes large in a very acidic medium by formation of AuO^+. As for silver, it has a very large zone of immunity, since the oxide Ag_2O is unstable at temperatures above about 200 °C. Silver remains almost free from weakly acidic media to basic media; it is substantially attacked by oxygen in very acidic media (formation of Ag^+) and in very basic media (formation of AgO^-) (Trémillon and Goret, 1967).

In other molten media, the situation can be very different, for example in alkali carbonates – where gold is found to be inoxidizable, platinum is made immune or passive by an oxide film, silver is corrodable with formation of Ag^+, and nickel is attacked with formation of

NiO not leading to passivity (Ingram and Janz, 1965) – or in molten alkali nitrates – in contact with which silver remains practically free from attack (above 200 °C), especially in presence of nitrite which plays the part of a reducing agent which completely immunizes the metal (Conte and Ingram, 1968).

This diversity of metal behaviour in molten media of different types constitutes one of the most attractive features of these media. The potential-acidity graphs, showing in diagram form the most important properties, are therefore of fundamental interest, in particular in connection with chemistry and electrochemistry in liquid media at high temperatures.

CHAPTER V

CORRELATION BETWEEN PROPERTIES IN DIFFERENT SOLVENTS

The subject of the last chapter forms the necessary complement for understanding and reasoned exploitation of non-aqueous media; this deals with the connection between chemical properties in the different solvents, the realization and analysis of which are essential elements for understanding of the role of the solvent.

We will consider in greater detail the relationship between acid-base properties, as defined and described in Chapter II. Realization and analysis of the correlation in this case are made more complete than for other types of property by the fact of a more complete knowledge of the acid-base properties in a fairly large number of different solvents.

The problem can be approached by relying on general thermodynamic considerations of solutions, which will form the subject of the first part of the chapter.

V.1. Thermodynamic Aspect of Solvation

Let us remember from the start that each solute, i, is characterized by a value of the 'chemical potential' μ_i, to which there corresponds an 'absolute activity' $\lambda_i = \exp(\mu_i/RT)$. In fact, only the chemical potential differences are accessible; we therefore define for a species in a given medium (that is in a given solvent) the 'relative activity' referred to a fixed 'standard' – corresponding to the chemical potential μ_i^0 and to the absolute activity λ_i^0 – by:

$$a_i = \frac{\lambda_i}{\lambda_i^0} = \exp \frac{\mu_i - \mu_i^0}{RT}. \qquad \text{(V-1)}$$

For solutions, the standard state is chosen in such a manner that, for an 'ideal' behaviour, the relative activity is regarded as equal to the concentration of the solute. If we apply a completely rigorous treatment, this ideal behaviour takes place only in an infinitely diluted solution, but in practice, a behaviour near ideality is reached when dilution is sufficient. Concentrated solutions deviate from ideality, which is allowed for by introduction of an 'activity coefficient', equal to the ratio between activity and concentration, tending to unity at infinite dilution. For ions in molecular solutions, an evaluation of ionic activity coefficients is made possible by the theory of Debye and Hückel (1923).

Solute transfer from one solvent to another

Solvation of a chemical species leads to energy consequences, which vary both with the nature of the solute and that of the solvent. As a result of this, at the same concentration in two different solvents* – more exactly, at the same value of the relative activity – a solute is characterized by different values of its chemical potential (or of its absolute activity). The chemical potential or absolute activity are smaller as solvation of the solute (or of its constituents if it is dissociated) takes place with increasing energy.

Having regard to the expression $\mu_i = \mu_i^0 + 2.3RT \log a_i$ (derived from V-1), the chemical potential difference existing between two solutions where the solute i has the same relative activity a_i (approximately at the same concentration c_i) is equal to the difference in standard chemical potentials μ_i^0 (value of the chemical potential when $a_i = 1$). This difference corresponds to the variation in free enthalpy due to the change in solvation when the solute (one mole) is transferred from one solvent to the other; this is called the free enthalpy of transfer of the solute from solvent R (reference solvent) to solvent S:

$$(\Delta G_t^0)_{R \to S} = (\mu_i^0)_S - (\mu_i^0)_R \,. \qquad \text{(V-2)}$$

The idea of free enthalpy of transfer enables us to find a correspondence between an activity $(a_i)_R$ of the solute in solvent R and an activity $(a_i)_S$ in solvent S such that the solute i has the same absolute activity in the two solutions. We have in fact:

$$(\mu_i^0)_R + 2.3RT \log (a_i)_R = \\ = (\mu_i^0)_S + 2.3RT \log (a_i)_S \,. \qquad \text{(V-3)}$$

* By different solvents, we mean liquids of different chemical constitution together with mixtures of solvents in variable proportions, such as water and alcohol, or water and dimethylformamide.

From which:

$$\log(a_i)_R - \log(a_i)_S = \frac{(\Delta G_t^0)_{R\to S}}{2.3RT}. \qquad \text{(V-4)}$$

Putting

$$\frac{(\Delta G_t^0)_{R\to S}}{2.3RT} = \gamma_t(i)_{R\to S} \qquad \text{(V-5)}$$

the expression (V-4) then takes the form:

$$(a_i)_R = \gamma_t(i)_{R\to S}\cdot(a_i)_S. \qquad \text{(V-6)}$$

γ_t is greater than unity or less than unity ($\log\gamma_t$ is positive or negative) depending on whether the dissolved substance is more strongly solvated in solvent R than in solvent S, or vice versa.

Several names have been suggested for the quantity γ_t so defined: distribution coefficient* (Kolthoff *et al.*, 1938), degenerate activity coefficient (Grunwald *et al.*, 1951), medium effect (Bates, 1964), solvation activity coefficient (Charlot and Trémillon, 1963; Vedel, 1967); we will use here the name recently suggested by the I.U.P.A.C. (1973): transfer activity coefficient. Various symbols have been suggested; γ_t has been adopted so as to stress the analogy with the activity coefficients mentioned

* In the case of partition of a molecular solute between an aqueous phase and another solvent S immiscible with water, the distribution coefficient of the solute – ratio of concentration in solvent S to concentration in aqueous phase –, in equilibrium (the same chemical potential of the solute in the two phases) and in sufficiently dilute solution, is equal to the reciprocal of γ_t.

at the start of the chapter, which are in current use.

In the case of solvents at ordinary temperatures (molecular solvents) and solutes which can exist in them as well as in aqueous solution, it is natural to choose water as a comparison solvent and to establish a correspondence between any relative activity of the solute in any solvent S and a relative activity in aqueous solution $(a_i)_{aq}$, such that the chemical potentials of the solute are identical. This correspondence is effected by means of the value of the free enthalpy of transfer of the solute from water to solvent S, to which there corresponds a value of the transfer activity coefficient $\gamma_t(i)_{\text{water}\to S}$:

$$(a_i)_{aq} = \gamma_t(i) \cdot (a_i)_S, \tag{V-7}$$

with:

$$\log \gamma_t(i) = \frac{(\Delta G_t^0)_{\text{water}\to S}}{2.3RT} = \frac{(\mu_i^0)_S - (\mu_i^0)_{\text{water}}}{2.3RT}. \tag{V-8}$$

Due to the expression (V-7) for the 'aqueous' activity, we can say that it is possible to express the activities of a solute in any solution on a common scale, for which the common reference state is that of the solute when infinitely diluted in aqueous solution, and the common standard state is that in aqueous solution.

Mean transfer activity coefficient of an electrolyte and individual transfer activity coefficients of its constituent ions

An ionic species in solution arises from the dissolving of

an electrolyte (ionophoric or ionogenic substance), which must give at least two different ions, an anion and a cation, and we cannot consider the transfer from one solvent to another of a single ionic species, but only the transfer of the anion and cation together. As a result of this fact, the value of the free enthalpy of transfer of one mole of the electrolyte is all that can be determined.

From this we can define a mean transfer activity coefficient $\gamma_{t\pm}$, in the same way in which we define the mean activity $a_{\pm}$ of the electrolyte; for a (strong) electrolyte of type $AB_n \rightarrow A + nB$ (without stating the charges on the ions A and B):

$$\log \gamma_{t\pm}(AB_n) = \frac{\Delta G_t^0(AB_n)}{2.3(n+1)RT}. \qquad (V\text{-}9)$$

This mean activity coefficient is the consequence of the mean of the solvation effects on the $n+1$ ions brought into solution by the electrolyte AB_n when completely dissociated.

We can attribute to each of the ions a part of the total free enthalpy of transfer of the electrolyte:

$$\Delta G_t^0(AB_n) = \Delta G_t^0(A) + n\Delta G_t^0(B). \qquad (V\text{-}10)$$

Each of these parts can be transformed into a transfer activity coefficient which will be attributed to each of the ions A and B individually, as in (V-8), leading to the relationship between the mean transfer activity coefficient of the electrolyte and the individual transfer activity coefficients of the constituent ions:

$$\gamma_{t\pm}(AB_n) = [\gamma_t(A) \cdot \gamma_t(B)^n]^{1/n+1}. \tag{V-11}$$

For a 1-1 electrolyte, in particular (type AB):

$$\gamma_{t\pm} = [\gamma_t(A) \cdot \gamma_t(B)]^{1/2}. \tag{V-12}$$

This relationship is similar to that existing between the mean activity $a_{\pm}$ and the activities of the constituent ions: $a_{\pm}=(a_A \cdot a_B)^{1/2}$.

ELECTROSTATIC EFFECT IN MOLECULAR SOLVENTS AND ITS CONTRIBUTION TO THE VALUES OF THE TRANSFER ACTIVITY COEFFICIENTS OF THE IONS AND OF THE MEAN TRANSFER ACTIVITY COEFFICIENTS OF THE ELECTROLYTES

The transfer of an ion, from one molecular solvent to another where the dielectric constant is different, is accompanied by a change in free enthalpy due to the change in electrostatic interactions (coulombic forces) between the ion and the dielectric medium. This energy change can be evaluated in an approximate manner by means of Born's equation (1920) – derived on the basis of a simple model of rigid spheres, uniformly charged, to which the ions are assimilated, and placed in a continuous dielectric medium, the solvent of dielectric constant ε.

For an ion of charge z and radius r, the electrostatic contribution to the free enthalpy of transfer, from water (dielectric constant $\varepsilon_{water}=78$ at ordinary temperature) to a solvent S (dielectric constant ε_S), is thus given by:

$$\Delta G^0_{t,\,\mathrm{el}} = 2.3RT \log \gamma_{t,\,\mathrm{el}} = \frac{Ne^2z^2}{2r}\left(\frac{1}{\varepsilon_S} - \frac{1}{\varepsilon_{\mathrm{water}}}\right) \tag{V-13}$$

(N=Avogadro's number, e=electron charge. The numerical value of $\frac{1}{2}Ne^2$ is about 165 Å kcal mol^{-1}).

According to this formula, the electrostatic effect must lead to an increase in the absolute activity when an ion is transferred from water to a solvent S of lower dielectric constant – which is the most frequent case.

Equation (V-13) also enables us to deduce the value of the electrostatic effect contribution to the mean transfer activity coefficient of an electrolyte. Thus for a 1-1 electrolyte consisting of cations of charge $+z$ and radius r_c and anions of charge $-z$ and radius r_A, we obtain:

$$\Delta G^0_{t,\,\mathrm{el}} = 2.3RT \log \gamma_{t,\,\pm\mathrm{el}} = \frac{Ne^2z^2}{2}\left(\frac{1}{\varepsilon_S} - \frac{1}{\varepsilon_{\mathrm{water}}}\right)\left(\frac{1}{r_A} + \frac{1}{r_C}\right). \tag{V-14}$$

We must note however that Born's equation can only provide a fairly rough approximation of the electrostatic contribution to the free enthalpy of transfer. Reality is in fact somewhat different from the ideal model postulated by this theoretical equation; the values of ionic radii, if we need to take solvated ions into consideration, vary appreciably with the nature of the solvent; in addition, it is certainly incorrect to treat solutions as if they were continuous dielectric media, having the same dielectric constant at all points (the so-called 'macroscopic' dielectric constant of the solvent), since the intense electric field operating at the surface of the ions and orienting the dipole molecules of the solvent, can cause changes in the dielectric constant in this region (dielectric saturation). We cannot therefore expect a very good verification of the variation law of $\log \gamma_t$ or $\log \gamma_{t\pm}$ inversely with the solvent dielectric constant.

In addition, other types of interaction (ion-dipole interactions, hydrogen bonds, even coordination bonds) play a part in solvation phenomena, at the same time that electrostatic interactions, and $\Delta G^0_{t,\,\mathrm{el}}$ constitute only a part of the total free enthalpy of transfer of an ionic solute, not necessarily a dominant part. We can form an estimate of the relative importances of the two parts $\Delta G^0_{t,\,\mathrm{el}}$ and $\Delta G^0_{t,\,\mathrm{non\ el}}$ (which are added together to determine the total value of ΔG_t^0) by establishing on one hand the importance the free enthalpy of transfer of molecular solutes (for which $\Delta G^0_{t,\,\mathrm{el}}=0$) can assume, and on the other hand the differences between $\Delta G^0_{t,\,\mathrm{el}}$ calculated from the formula (V-14) for the transfer of an electrolyte and the total values of ΔG_t^0 as determined experimentally under the same conditions, differences that cannot be explained by uncertainty about the results of the calculations. For example, for the transfer of hydrochloric acid from water to the following two hydro-organic mixtures, it was obtained (from Bates, 1968):

	ΔG_t^0 measured	$\Delta G_{t,\,el}^{0}$ calculated (with $r_{Cl^-} = 4.6$ Å and $r_{H^+} = 2.8$ Å)
90wt % methanol	1.9 kcal mol^{-1}	1.3 kcal mol^{-1}
70wt % dioxane	2.6 kcal mol^{-1}	4.1 kcal mol^{-1}

The energy $\Delta G^0_{t,\,\mathrm{non\ el}}$ corresponding to non-coulombic interactions, more or less specific, cannot be evaluated by means of mathematical formulae as easily as the coulombic part, in the present state of physico-chemical theories. In this situation, the possibilities of prediction of transfer activity coefficients are limited, and in consequence recourse must be made to experiments in order to evaluate them.

Nevertheless, the laws derived from Born's equation provide very useful guidance in interpretation or rough prediction of modifications to chemical properties resulting from changes in the particular factor which is the dielectric constant of a molecular solvent.

Experimental evaluation of transfer activity coefficients

Several methods have been suggested for determination of the transfer activity coefficient values of various solutes in molecular solvents.

1. In the case of molecular substances (non-electrolytes) which dissolve in water only in the state of molecules

(without dissociation or combination with species other than solvent molecules), the experimental method used consists of solubility measurements; either under a determined partial pressure for a volatile substance (dissolved in a solvent much less volatile), or by saturation of the solvent with the pure substance when this latter is a solid or a liquid of low solubility (solubility is the concentration of the saturated solution).

We can say that the solubility $\mathcal{S}$ becomes larger as the molecules of dissolved substance are more strongly solvated; solubility therefore varies with the solvent. Since the chemical potential of the substance in all saturated solutions, in equilibrium with the pure substance, is the same and equal to that of the latter, it appears that its transfer activity coefficient is equal to the ratio between its relative activity in saturated aqueous solution to its relative activity in the saturated solution in solvent S. If the solubilities $\mathcal{S}_{\text{water}}$ and $\mathcal{S}_S$ are sufficiently low for the relative activities to be approximately equal to them, the transfer activity coefficient is more simply the ratio between the values of these solubilities:

$$\gamma_t = \frac{\mathcal{S}_{\text{water}}}{\mathcal{S}_S}. \qquad \text{(V-15)}$$

When the solvent S is immiscible with water, we can still imagine measurements of distribution of the substance between the two liquid phases, the transfer activity coefficient then corresponding to the reciprocal of the distribution coefficient $D = a_S / a_{\text{water}}$. We can in this case operate under conditions of dilution for which the relative activity may be regarded as practically equal to the concentration, which is more easily determined, even if the substance has a high solubility.

2. In the case of strong electrolytes, which dissolve exclusively in the state of free solvated ions (without intervention of combinations with species of the solvent or of the dissolving solution, such as solvolysis products or complexes), it is still possible to make use of solubility measurements, under the same conditions as above. The ratio of the solubilities in water and in solvent S, when these are sufficiently low, leads in this case to the value of the mean transfer activity coefficient of the substance; in the case of a 1-1 electrolyte for example:

$$\gamma_{t\pm} = (\gamma_{t+} \cdot \gamma_{t-})^{1/2} = \frac{\mathscr{S}_{\text{water}}}{\mathscr{S}_S}. \tag{V-16}$$

This relation follows also from the fact that the electrolyte, in particular a salt, has the same chemical potential (the sum of the chemical potential of each ion) in the two saturated solutions, equal to that of the pure substance in equilibrium with any saturated solution. It ceases to be valid however if the saturated solutions are each in equilibrium with a different 'solvate' of the salt. For example, a saturated aqueous solution of sodium iodide is in equilibrium with the solid hydrate $NaI, 2H_2O$, whilst a saturated solution in acetone is in equilibrium with the solid solvate $NaI, 3CH_3COCH_3$; the two saturated solutions are in equilibrium neither with each other nor with the pure salt NaI, and the relationship (V-16) cannot be used.

Another method used for determination of mean transfer activity coefficients is measurement of electromotive forces of cells using the electrolytic solutions under consideration. For example, the difference between the standard electromotive forces E^0_{water} and E^0_S of the two similar cells:

$(Pt)H_2(1\ \text{atm})$ (hydrogen electrode)	$H^+X^-(a_\pm=1)$ in water or in solvent S (electrolyte)	AgX(solid), Ag (reference electrode)

depends only on the free enthalpy of transfer of the strong acid $HX(X^-=Cl^-, Br^-$ or $I^-)$ from water to solvent S, since all the other constituents are maintained at the same chemical potential. So, we obtain:

$$\Delta G_t^0(\text{HX}) = 4.6RT \log \gamma_{t\pm}(\text{HX}) = F(E^0_{\text{water}} - E^0_S). \tag{V-17}$$

In the same manner, by replacing the hydrogen electrode by an electrode of an amalgam of a metal M, and H^+X^- by the ionized salt M^+X^- in solution, either in water or in solvent S, the difference between the standard electromotive forces leads to the mean transfer activity coefficient of M^+X^-.

EVALUATION OF TRANSFER ACTIVITY COEFFICIENTS OF INDIVIDUAL IONS

It has been shown that it is impossible thermodynamically to attain the value of the transfer activity coefficient of an individual ion. Indeed, transfer of a single ion from one solvent to another implies a transfer of electric charge, which requires electrical work to be done; this latter is to be added to the solvation energy of the ion. It depends on the difference of internal electric potential between the two phases considered, when brought into contact, a difference which remains inaccessible to experiments.

In order to render an account of this phenomenon, Guggenheim (1929) characterized the thermodynamic state of an ionic species i of charge z_i, in a given homogeneous medium, by its 'electrochemical potential' $\tilde{\mu}_i$, related to its chemical by $\tilde{\mu}_i = \mu_i + z_i F \phi$, where ϕ is the internal electric potential of the medium. In this manner, equilibrium of an ionic species between two different solvents R and S implies equality of the values, not of its chemical potential, but of its electrochemical potential in the two phases in contact; in place of the relation (V-4), we then have:

$$\log(a_i)_R - \log(a_i)_S = \log \gamma_t(i)_{R \to S} + z_i F(\phi_S - \phi_R). \tag{V-18}$$

The internal potential difference $\phi_S - \phi_R$, which controls the ratio between the relative activities of the ion in the two solvents, is known as the 'junction potential' E_j between the two solutions.

It is found impossible to separate the term of transfer activity coefficient (variation of the solvation effects) from the term due to the junction potential, when electrochemical measurements are used in view of determining the activity variations of a single ion. For example, let us consider the cells mentioned in the previous paragraph, with hydrogen electrode; the difference between the standard electromotive forces, $E^0_{\text{water}}—E^0_S$, is in fact equal to the electromotive force ΔE^0 of the cell, with a liquid junction, constituted of two normal hydrogen electrodes, one in aqueous solution and the other in solvent S:

(1) (Pt)H_2(1 atm)	$H^+X^-(a=1)$ in water	$H^+X^-(a=1)$ in S	(Pt)H_2(1 atm) (2)

Evaluation of this electromotive force, using the activities of H^+ ion only, is effected as follows.

Electrochemical equilibrium ($H^+ + e^- \rightleftharpoons \frac{1}{2}H_2$), at each of the platinum electrodes, is expressed by:

$$(\tilde{\mu}_{e^-})_{\text{Pt1}} = \tfrac{1}{2}\mu^0_{H_2} - (\tilde{\mu}^0_{H^+})_{\text{water}} \tag{V-19}$$

$$(\tilde{\mu}_{e^-})_{\text{Pt2}} = \tfrac{1}{2}\mu^0_{H_2} - (\tilde{\mu}^0_{H^+})_S\,. \tag{V-20}$$

Remembering that $(\tilde{\mu}_{e^-})_{\text{Pt1}} = \mu_{e^-} - F\phi_{\text{Pt1}}$ and that $(\tilde{\mu}_{e^-})_{\text{Pt2}} = \mu_{e^-} - F\phi_{\text{Pt2}}$, we have:

$$\Delta E^0 = \phi_{\text{Pt2}} - \phi_{\text{Pt1}} = \frac{1}{F}\,[(\tilde{\mu}^0_{H^+})_S - (\tilde{\mu}^0_{H^+})_{\text{water}}]$$

$$= \frac{1}{F}[(\mu^0_{\mathrm{H}^+})_S - (\mu^0_{\mathrm{H}^+})_{\mathrm{water}}] + (\phi_S - \phi_{\mathrm{water}}). \tag{V-21}$$

Let:

$$\Delta E^0 = \frac{2.3RT}{F}\log\gamma_t(\mathrm{H}^+) + E_j. \tag{V-22}$$

Measurement of the electromotive force of the cell under discussion, constituted of two half cells (with liquid junction) each comprising a different solvent, only enables us to determine the difference in standard electrochemical potentials, and not that of the standard chemical potentials of the individual ion, since the junction potential between the two different solvents is unknown.

The same argument applies for any system using an ion other than H^+.

Experimental measurements do not therefore lead to the values of the transfer activity coefficients of single ions but only to the values of the mean transfer activity coefficients of two ions of opposite sign; hence we only know values of the products of transfer activity coefficients of a cation and an anion (or by combination, values of the ratio of the coefficients for two cations or for two anions).

It would suffice if we could evaluate one single transfer activity coefficient, so as to deduce from it without difficulty all the values for the other ions, starting from the mean coefficients. But this starting point can only be established by adopting 'extra-thermodynamic' hypotheses about solvation of one or several reference ions; these hypotheses cannot be experimentally controlled, and therefore are subject to uncertainty. However, comparison of the results obtained by several different methods provides a means of estimation of their accuracy, depending on whether agreement is good or not. These last years have seen an increased effort towards a solution of this problem. A brief analysis of the methods used is of interest in order to bring out the methods of thought adopted in this area, together with concrete working hypotheses.

1. The ideal would be to have available a reference ion having zero free enthalpy of transfer in all solvents. Such an ion does not exist, but we can assume that we approach closely to this for an ion of which the energy of solvation is as small as possible in all solvents. For this to be the case, the ion must be as large as possible, with the smallest possible charge (therefore $z=1$), and with a low polarizability, which leads to choice of a cation. Pleskov (1947) suggested that the rubidium Rb^+ ion is the one which best meets these conditions; the caesium ion Cs^+, still larger, was not used because there was only a small amount of information about it, but it has later been proved that the difference in behaviour between these two ions is quite small.

If we assume the hypothesis of a transfer activity coefficient equal to unity for the cation Rb^+

$$[\Delta G_t^0(\mathrm{Rb}^+) = 2.3RT \log \gamma_t(\mathrm{Rb}^+) = 0],$$

it becomes possible to evaluate the other ionic transfer activity coefficients, either starting from measurements of solubility or of distribution of rubidium salts — since $\gamma_{t-} = \gamma_{t\pm}^2$ –, or starting from measurements of electromotive forces of cells. In this latter case, considering the rubidium amalgam/Rb^+ electrode as retaining a standard potential invariable whatever solvent is used – since the chemical potentials of the metal and of the cation remain unaltered –, the change in electromotive force of a cell such as:

Rb(amalgam)	$Rb^+Cl^-(a=1)$ in water or in solvent S	AgCl(solid);Ag

will correspond to the change in chemical potential of the anion Cl^-, the only species in the cell the chemical potential of which changes with a change in solvent. In the same way, the change in electromotive force of a cell of the following type, where M represents any metal whatsoever:

Rb(amalgam)	$Rb^+(a=1)$ in water or in solvent S	$M^{z+}(a=1)$ in the same solvent	M

will correspond – with a correction for the junction potential, which is small in this case since the two compartments of the cell contain the same solvent – to the change in chemical potential of the single cation M^{z+} and will provide values of the transfer activity coefficient of this latter. For the hydrogen ion, the values of the transfer activity coefficient are obtained in the same way by use of a hydrogen electrode in place of the M/M^{z+} system.

Pleskov's hypothesis is only a first approximation, and we can contemplate an improvement paying regard to the electrostatic contribution, which is small but easily calculable, to the free enthalpy of transfer of the Rb^+ ion (of which it must form the greater part). This correction was introduced by Strehlow (1952, 1966) *et al.* (1960), who made use of Born's formula with a refinement allowing for the possible change of the radius of the solvated ion during transfer; this leads to the expression following from formula (V-13):

$$\Delta G^0_{t,\,\mathrm{el}}(\mathrm{Rb}^+) = 165\left[\frac{1 - 1/78}{1.5 + 0.85} - \frac{1 - 1/\varepsilon_S}{1.5 + r'}\right] \qquad \text{(V-23)}$$

1.5 is the crystallographic radius in Å of the Rb^+ ion, 0.85 and r' the values in Å found as the increase in ionic radius to allow for solvation in water in the first case and in solvent S in the second case. This correction is found to be small and only confirms the method suggested by Pleskov for evalution of the transfer activity coefficients of single ions.

2. But potentiometric measurements with a rubidium electrode are difficult to carry out, and are also subject to a large risk of error, if not made impossible, because of attack of the alkali metal by the solvent (especially in the case of 'acidic' media).

Strehlow *et al.* (1960) therefore suggested and justified the replacement of the Rb/Rb^+ couple by another oxidizing-reducing couple technically more suited and capable of playing the same part of a potentiometric reference during measurements of electromotive forces on change of solvent.

The only theoretical condition to be satisfied for the validity of this assumption is that the ratio of the absolute activities of the two components of the couple is preserved during all changes of solvent. This implies firstly that these components both exist in solution (unsaturated). Further, since there must be a difference in electrical charge between them, at least one is ionic; its free enthalpy of transfer comprises an electrostatic part not found in the other, and which varies with changes in the dielectric constant. It then becomes necessary, in order to satisfy the condition imposed, to reduce to the minimum this electrostatic part, by adoption of the following criteria:

the electric charges to be as small as possible, hence a molecular reductant and a monovalent cationic oxidant (system $Ox^{+} + e^{-} = R$);

species practically spherical in shape and with as large a radius as possible, hence complex species which do not change in structure on electron exchange.

It is not easy to find such oxidizing-reducing couples which are satisfactory in practice, since complex species of large radii are liable not to have rapid electron exchanges, and so do not allow measurements of equilibrium potential; further, a large complex molecule (probably organic) runs the risk of not being solvated by water and of not being sufficiently soluble in it. Nevertheless, out of eighteen systems tested with this end in view, Strehlow and his co-workers retained two which appeared to be suitable: the couples ferrocene-ferricinium$^+$ cation and cobaltocene-cobalticinium$^+$ cation. Ferrocene is a molecular complex of sandwich structure, formed from a ferrous cation Fe^{2+} and two cyclopentadienyl anions (by hybridation of the π orbitals of the dienic cycles with the *sd* orbitals of the ferrous ions); it is almost spherical and has a mean radius of 2.3 Å. It can lose an electron, leading to the ferricinium cation, which is identical in structure and has a symmetrically distributed positive charge. Cobaltocene has almost identical properties (Co^{2+} replacing Fe^{2+}); it is however appreciably more strongly reducing. The solubilities of ferrocene and of cobaltocene in water are small but adequate (3×10^{-5} mol l^{-1} for ferrocene); they are much greater in organic solvents (for example, ferrocene in ethanol: 0.12; in acetone: 0.28; in tetrahydrofuran: 1.1 mol kg^{-1}). On the other hand, ferricinium and cobalticinium salts are very soluble in water but less so in organic solvents – this being perhaps due to the solvation effect of the counter- ion. The standard potentials for the two couples (0.40 V for ferrocene and —0.92 V for cobaltocene, with respect to the normal hydrogen electrode in water) are much better 'centred' than that of the couple Rb^+/Rb amalgam.

Experimental measurements are carried out in the same manner as with the couple Rb/Rb^+, replacing the half-cell

Rb(amalgam) | Rb^+ (a=1) in water or in S ||

by:

Pt or Hg | ferrocene+ferricinium$^+$ in water or in S||

By associating this half-cell with a second half-cell such as:

|| M^+ (a=1) in the same solvent | M

the difference in standard electromotive forces ΔE^0 (after correction for the liquid junction potential) corresponds to the sum of the free enthalpies of transfer:

$$F \cdot \Delta E^0 = \Delta G_t^0(\mathrm{M}^+) + \\ + [\Delta G_t(\text{ferrocene}) - \Delta G_t(\text{ferricinium})]. \qquad \text{(V-24)}$$

In so far as the term in brackets remains practically zero, in accordance with the initial hypothesis, measurement of ΔE^0 enables us to find the value of:

$$\Delta G_t^0(\mathrm{M}^+) = 2.3RT \log \gamma_t(\mathrm{M}^+).$$

Independently of this hypothesis, Strehlow defined a function which he called R_0, starting from the measured values of the differences in electromotive force, and which becomes under standard conditions:

$$R_0^0(\mathrm{M}^+) = -\frac{F \cdot \Delta E^0}{2.3RT}. \qquad \text{(V-25)}$$

If we assume the hypothesis which permits us to neglect the term in brackets in expression (V-24), this experimental quantity then becomes nearly equal to the negative logarithm of the transfer activity coefficient:

$$\log \gamma_t(\mathrm{M}^+) \approx - R_0^0(\mathrm{M}^+). \tag{V-26}$$

An argument in favour of this method of evaluation of ionic transfer activity coefficients is the good agreement observed in several solvents between the results obtained from couples of ferrocene, cobaltocene and rubidium, their differences of standard potential being preserved within experimental error limits. It would appear justified to apply a correction for electrostatic effect to the ferricinium and cobalticinium ions, identical to that applied to the rubidium ion, but this effect is even smaller than in the case of rubidium, and can safely be neglected.

Nelson and Iwamoto (1961) suggested another reference oxidizing-reducing couple, consisting of the ferrous and ferric complexes of 4,7-dimethyl 1,10-phenanthroline. Although the results obtained with this system are fairly close to those with the rubidium method, the high charges of the complexes (+2 and +3) makes it a priori less suitable than the two other systems.

3. Apart from potentiometric determinations, a different method of evaluation based on a particularly simple hypothesis, stated by Grunwald *et al.* (1958–1962), has been applied by Popovych and Dill (1966–1969) and also by Parker *et al.* (1966–1967). This hypothesis postulates that certain ions, cations or anions, undergo the same change in chemical potential on transfer from water to another solvent; equal transfer activity coefficients are therefore

assumed for these particular ions. The ions of this type suggested by Grunwald are tetraphenylarsonium ϕ_4As^+ or tetraphenylphosphonium ϕ_4P^+ and tetraphenylborate ϕ_4B^-; Popovych added the triisoamylbutylammonium ion. These species are effectively of the same almost spherical shape and of the same size; the first three comprise different central atoms surrounded in the same way by four groupings of large volume which isolate them from the solvent and give the three ions about the same specific interactions*.

Popovych and Parker took up this hypothesis to make use of solubility measurements of sparingly soluble salts both in water and in organic solvents. They assume, in one case for triisoamylbutylammonium tetraphenylborate and in the other case for tetraphenylarsonium tetraphenylborate, that $\gamma_{t+} = \gamma_{t-} = \gamma_{t\pm}$. In consequence, the solubilities of tetraphenylarsonium iodide or of silver tetraphenylborate will provide, once we know the transfer activity coefficients of the tetraphenylarsonium and tetraphenylborate ions, those of the I^- or Ag^+ ions, etc. Thus all the values desired are obtainable one after the other.

4. One last interesting type of evaluation method consists of what might be called extrapolation to infinite ionic

* We can add that the free enthalpy of transfer of these ions must therefore be approximately equal to that (obtainable by experiment) of the molecular species tetraphenylmethane ϕ_4C or tetraphenyltin ϕ_4Sn, increased by the term for the electrostatic effect.

radius. This method uses the assumption that the free enthalpy of transfer tends to zero as the ionic radius increases to infinity. Using Born's equation as a base, Izmailov *et al.* (1958–1960), and later Feakins *et al.* (1963–1966) made graphs of the free enthalpy of transfer (for a solvent of given dielectric constant) of a series of strong electrolytes having a common ion – such as alkali chlorides MCl or halogen acids HX – as a function of the reciprocal of the radius r of the variable counter-ion, in the expectation of a straight line graph. Extrapolation to $1/r=0$, leading to a value of the free enthalpy of transfer of the variable ion which can be taken as zero, then gives that of the common ion of the series [$\Delta G_t^0(Cl^-)$ for the MCl series, or $\Delta G_t^0(H^+)$ for the HX series.]. In practice, if the experimental points lie on a good enough straight line, the lines obtained do not have the theoretical slope deduced from Born's equation, which casts doubts on the accuracy of the procedure.

Alfenaar and Deligny (1965–1967) have subsequently refined and improved the procedure by seeking to allow for the non-electrostatic part of the free enthalpy of transfer, which is not taken into account in Born's equation. Before graphing the values of free enthalpy of transfer versus $1/r$, they deduce from these a part called neutral, estimated to be equal to the free enthalpy of transfer of an uncharged species (a rare gas, for example) of the same radius as the variable counter-ion. This treatment appears more valid for large ions than for small ones, such that the change in $\Delta G_t^0 - \Delta G_{t,\,\text{neutral}}^0$, observed as a function of

$1/r$, must now tend towards a straight line of suitable slope as $1/r$ tends to zero. For small ions, due to the increased ion-dipole interactions, the line becomes curved when $1/r$ increases. In order to ensure suitable extrapolation, based on the experimental values of ΔG_t^0 for MX and for HX, Alfenaar and Deligny, drew graphs at the same time of the values of $\Delta G_t^0(H^+)+\Delta G_t^0(X^-)-\Delta G_{t,\,\text{neutral}}^0(X^-)$ and of $\Delta G_t^0(H^+)-\Delta G_t^0(M^+)+\Delta G_{t,\,\text{neutral}}^0(M^+)$ as a function of $1/r_{X^-}$ or $1/r_{M^+}$. Extrapolation of the two curves must lead to the same value of $\Delta G_t^0(M^+)$; in addition, their slopes at the origin must have the value given by Born's equation.

By way of conclusion on this topic, it does not appear reasonable at present to make a judgment as to the comparative validity of these different methods, owing to the too small number of results available. Nevertheless, we can say that the rubidium method appears the most justified for theoretical reasons, and that its excellent agreement with the ferrocene method probably makes these the two most reliable methods. We can further take note of the order of magnitude of the divergences found between the results obtained by the ferrocene method and the solubility method, in accordance with the values collected in table IX (which includes some examples only). Only by theoretical refinements, or consideration of a fairly large number of results obtained by the most diverse methods, will it be possible to evaluate the transfer activity coefficients of individual ions more closely, and thence to obtain more

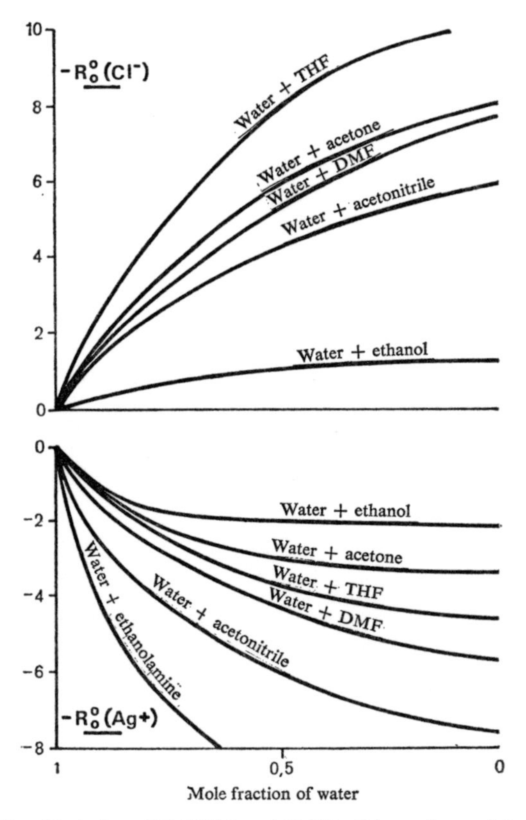

Fig. 11. Variation of $R_0{}^0(Cl^-)$ and $R_0{}^0(Ag^+)$ in various mixtures of water and organic solvents (according to Trémillon *et al.* 1968)

TABLI

Approximate experimental values of transfer activity coefficients ($\log \gamma_t$
and in acetamide at 98 °C (compared with aqueous solutions at the sam

Solvents	Cations			
	Ag^+	Cs^+	K^+	Φ_4As^+
Acetamide (98 °C)	—4.5 [3]			
Acetone	—3.4 [3]			
Acetonitrile	—5.6 [2]	—0.5 [2]	—0.4 [2]	—6.2 [2]
	—7.6 [3]			
DMF	—4.6 [2]	—2.4 [2]	—2.5 [2]	—7.0 [2]
	—5.7 [3]			
DMSO	—7.3 [2]		—3.3 [2]	—7.3 [2]
Ethanol	0.8 [2]	—0.1 [1]	0.0 [1]	—4.6 [2]
	—2.1 [3]	0.9 [2]	1.3 [2]	
HMPA	—7.6 [2]		—1.9 [2]	—7.8 [2]
THF	4.6 [3]			

Numerical values correspond to the activities expressed in the scale c
ments and extrapolation to infinite ionic radius (according to Strehlow)
(Parker *et al.*); [3] potentiometric method and hypothesis $\gamma_t(R) = \gamma_t(Ox)$ fc

reliable results for application purposes.

Certain rules emerge at the present time from an examination of the results available. In particular, it appears that many cations, if of small charge, are more strongly solvated by certain common dipolar organic solvents such as acetone, DMF, DMSO, acetonitrile, HMPA or THF than by water ($\log \gamma_t$ is negative). This is notably true for

X

f some cations and anions, in various solvents at ordinary temperature emperature).

Anions						
Φ_4B^-	Cl^-	Br^-	I^-	SCN^-	ClO_4^-	$CH_3CO_2^-$
	3.0 [3]	2.8 [3]	2.5 [3]	1.7 [3]		
	8.1 [3]	7.6 [3]	5.8 [3]	5.0 [3]		
—6.2 [2]	7.9 [2]	5.4 [2]	3.0	2.9 [2]		10.3 [2]
	5.9 [3]	6.0	4.1 [3]	4.3 [3]		
—7.0 [2]	8.3 [2]	6.3 [2]	3.4 [2]	3.2 [2]	—0.3 [2]	11.4 [2]
	7.7 [3]	6.1 [3]	3.1 [3]	3.2 [3]		
—7.3 [2]	6.7 [2]	4.4 [2]	1.5 [2]	1.3	—0.3 [2]	8.1 [2]
—4.6 [2]	4.1 [1]	3.6 [1]	3.0 [1]	0.2 [2]	0.1 [2]	1.9 [2]
	1.5 [2]	1.1 [2]	0.5 [2]			
			4.3 [3]			
—7.8 [2]	7.7 [2]	5.6 [2]		1.0 [2]		
	10.1 [3]	9.9 [3]	8.5 [3]	8.6 [3]		

nole fractions. Methods of determination used: [1] solubility measure- measurements of solubility products and hypothesis $\gamma_{t+} = \gamma_{t-}$ for $\Phi_4AsB\Phi_4$ errocene-ferricinium (according to Trémillon *et al.*).

Ag^+ ions for example. Figure 11 shows the variation of:

$$-R_0^0(Ag^+) \approx \log \gamma_t(Ag^+)$$

in some hydro-organic mixtures, as the solvent becomes richer in its organic constituent; ethanolamine appears to be the solvent which solvates the Ag^+ ion the most strongly, which is justified by the existence of coordination complexes between this cation and the nitrogenous organic molecules.

On the other hand, simple anions are more solvated by water than by aprotonic dipolar solvents, especially if they are small. The existence of hydrogen bonds between the water molecules and the small ions would appear to be the cause of this property, which is mainly seen in the series of halide anions, for which $\log \gamma_t$ is positive and increases in the order I^- ($< SCN^-$) $< Br^- < Cl^- < F^-$; the influence of the solvent is seen in Figure 11 for example, which shows the variations in $-R_0^0(Cl^-) \approx \log \gamma_t(Cl^-)$ in several hydro-organic mixtures. Nevertheless, large anions which are more strongly polarizable (such as Φ_4B^-, I^-, etc.) might appear, on the contrary, more solvated by aprotonic dipolar solvents than by water (Parker *et al.*, 1966–1967).

The greatest research effort has been devoted to solvation of the hydrogen ion, by reason of its practical importance and its particular behaviour, and these investigations have been mainly carried out in mixtures of water and another completely miscible solvent. The results obtained, such as those shown on figure 12, have shown up structural effects, both on the solvent (to explain the minimum of chemical potential of the H^+ ion solvated in mixtures of intermediate composition), or on the solvated species (preferential solvation by water, shown by the variation of $\log \gamma_t$ as a function of the activity of water in the mixture, or on the other hand intervention of molecules of organic

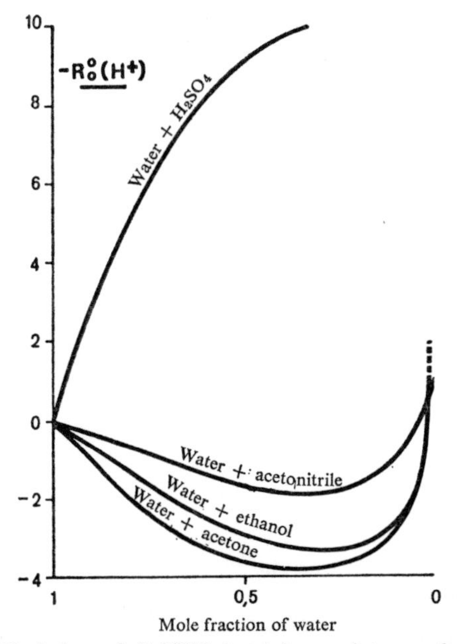

Fig. 12. Variation of $R_0{}^0(H^+)$ in various mixtures of water and another solvent (from Strehlow and Wendt, 1961, for water + H_2SO_4; Vedel, 1967).

solvent in place of water molecules in the primary ion solvation).

Relation between the values of the constant for the same chemical equilibrium in different solvents

Let us consider a chemical equilibrium which can be established in various solvents.

We can in a general way symbolize it by:

$$\sum \nu_A A \rightleftharpoons \sum \nu_B B \tag{V-27}$$

where A represents the species taking part in one of the reactions, B those taking part in the opposite reaction, and ν the respective stoichiometric coefficients.

Whatever the solvent, the equilibrium condition is expressed thermodynamically by:

$$\sum \nu_A \mu_A = \sum \nu_B \mu_B \,. \tag{V-28}$$

Having regard to the expression for chemical potentials, (V=28) leads to the relation expressing the mass action law, bringing in the relative activities of the species taking part*

$$\sum \nu_B \log a_B - \sum \nu_A \log a_A = \log K \,. \tag{V-29}$$

The value of the equilibrium constant $\log K$ is determined by the standard chemical potentials of the various constituents A and B. For this reason, the value is different according to the solvent in which we are considering establishment of equilibrium.

We therefore find thermodynamic justification of the experimental fact that one chemical equilibrium has different values for its constant depending on the solvent. The differences of solvation of the chemical species taking part lead to a shift of equilibrium.

* With the exception, of course, of the activities which remain unchanged: substances in saturated solution, that is in presence of the pure liquid or solid substances, or volatile substances under a constant partial pressure.

If $K(S)$ denotes the value of the constant in a solvent S and K(aq) its value in aqueous solution, relations (V-7) and (V-29) give the relationship between these two values:

$$\Delta(\log K) = \log K(S) - \log K(\text{aq}) = = \sum \nu_{\text{A}} \log \gamma_t(\text{A}) - \sum \nu_{\text{B}} \log \gamma_t(\text{B}). \quad \text{(V-30)}$$

Thus we can predict the modification in equilibrium constant when we pass from aqueous solutions to any solvent S whatsoever, from the values of the transfer activity coefficients of the different dissolved species taking part in the equilibrium under consideration.

Let us consider in particular the partial dissociation of a dissolved compound, of type AB_n, into two solvated constituents A and B:

$$AB_n \rightleftharpoons A + nB. \quad \text{(V-31)}$$

The value of the dissociation constant in a solvent S, compared with its value in aqueous solution, is given by:

$$\Delta(\log K) = \log \gamma_t(AB_n) - \log \gamma_t(A) - n \log \gamma_t(B) \quad \text{(V-32)}$$

Thus, when for example the increase in solvation of A and of B is greater than that of the compound AB_n, or also when the reduction in solvation of the compound is greater than that of the total products of dissociation, the constant $K(S)$ is greater than K(aq): dissociation takes place to a greater extent in S than in water. It is therefore necessary to have regard to changes in solvation both of the compound and of the dissociation products, in order to interpret or predict the change in dissociation equilibrium. This entirely general reasoning applies equally well to dissociation of acids as to that of complexes in solution.

In the case of equilibrium of dissolution of a sparingly soluble salt of formula $A_\alpha B_\beta$ and producing in solution α cations of A for β anions of B, the solubility product $K_S = a_A^\alpha \cdot a_B^\beta$ does not have the same value in aqueous solution as in solvent S and depends on the solvation of the A and B ions; the values of K_S are connected by:

$$\Delta(\log K_s) = \alpha \log \gamma_t(A) + \beta \log \gamma_t(B). \tag{V-33}$$

Electrostatic contribution to the variation in equilibrium constant from one molecular solvent to another

By virtue of what has previously been said about the electrostatic contribution to the free enthalpy of transfer of ions, it is clear that the single action of modifying the dielectric constant must lead to a change in the constant of an ionic equilibrium and hence to a shift of this latter, quite apart from the effect of other more specific types of interaction. The change is different, and greater or less in size, according to the number and the nature of ions which take part in each of the opposing reactions; this leads to a distinction between several types of ionic equilibrium (equilibria between molecules are clearly a priori unaffected by changes in dielectric constant).

(a) An equilibrium which has the same number of ions of the same charge in the two opposing reactions is said to be 'iso-ionic'. In particular, in the simplest case:

$$AB^+ \rightleftharpoons A^+ + B \qquad \text{(V-34)}$$

we can put, in accordance with relations (V-13) and (V-35):

$$\Delta (\log K)_{el} = k \left(\frac{1}{r_{AB^+}} - \frac{1}{r_{A^+}} \right) \cdot \Delta \left(\frac{1}{\varepsilon} \right). \qquad \text{(V-35)}$$

If r_{A^+} and r_{AB^+} remain of the same order of magnitude, the change in log K will be practically zero; we can regard an iso-ionic equilibrium as being practically unaffected by changes in the solvent dielectric constant (but it can be affected by changes in other factors).

This is the case, for example, in dissociation of cationic acids of type HA^+, like NH_4^+, HR_3N^+, etc. ($HA^+ \rightleftharpoons A + H^+$), of metallic ion complexes with molecular ligands – ammines, in particular: $M(NH_3)_n^{z+} \rightleftharpoons M^{z+} + nNH_3$ – or also of complexes such as I_3^- ($I_3^- \rightleftharpoons I_2 + I^-$), although the appreciable difference in size between the two anions I_3^- and I^- can bring about an appreciable electrostatic effect on the dissociation, a fall in dielectric constant being favourable to the largest ion, and hence unfavourable to dissociation*.

(b) On the other hand, the dissociation constant of a

* From experiment, it appears that the stability of I_3^- is larger in solvents of low dielectric constant than in water, despite the large increase in solubility of iodine which should favour dissociation.

compound which produces a larger number of electric charges must, in principle, be much more sensitive to changes in the solvent dielectric constant; if all other things are equal (for the other effects of solvation), $\log K$ must vary inversely with the dielectric constant. In particular, in the simplest case:

$$AB \rightleftharpoons A^+ + B^- \tag{V-36}$$

$$\Delta (\log K)_{el} = -k \left(\frac{1}{r_{A^+}} + \frac{1}{r_{B^-}} \right) \cdot \Delta \left(\frac{1}{\varepsilon} \right). \tag{V-37}$$

This means that dissociation becomes less when the dielectric constant of the solvent falls.

In the same way, for an equilibrium of type:

$$AB^+ \rightleftharpoons A^{2+} + B^- \tag{V-38}$$

$$\Delta (\log K)_{el} = -k \left(\frac{4}{r_{A^{2+}}} + \frac{1}{r_{B^-}} - \frac{1}{r_{AB^+}} \right) \cdot \Delta \left(\frac{1}{\varepsilon} \right). \tag{V-39}$$

Assuming that r_{A^+}, r_{B^-} and r_{AB^+} remain of the same order of magnitude, it appears that the variation of log K as a function of $1/\varepsilon$ in the case of equilibrium (V-38), is theoretically about twice as fast as that in the case of (V-36).

The coefficient $k = Ne^2/4.6RT$ has an approximate value of 120 at 20 °C, when the ionic radii are expressed in Ångstroms. In the above expressions, we must remember that it is postulated that the ionic radii are not modified by the change in solvent, which is clearly only a rough approximation.

This influence of variations in dielectric constant of the medium can in principle be applied equally well to the molecular acids (of type $HA \rightleftharpoons A^- + H^+$) and anionic acids (of type $HA^{n-} \rightleftharpoons A^{(n+1)-} + H^+$) as to complexes between metallic cations and anionic ligands. In this way, we must find in general that when we change from one solvent to another of lower dielectric constant, there is a weakening of the molecular and anionic acids with respect to the cationic ones, and an increase in stability of the complexes with anionic ligands, with respect to complexes with molecular ones. In reality, since the electrostatic effect enters only into one part of the modification of the equilibrium constant, the laws to which it leads are only rarely obeyed.

It is true that a variation of $\log K$ inversely with the dielectric constant has sometimes been observed, in mixtures of miscible solvents in varying proportions (hydro-organic mixtures mainly), for acid-base equilibria or complex formation equilibria, showing in these cases that the electrostatic effect is preponderant – or possibly that the other effects lead to a similar variation. This effect is best observed in mixtures of water and alcohols, water and dioxane, or water and acetone. For example, the formation constants, $\log K_1$, $\log K_2$, $\log K_3$, $\log K_4$ of the complexes $CdCl^+$, $CdCl_2$, $CdCl_3$ and $CdCl_4^{2-}$, in mixtures of water and methanol (0 to 80%), are effectively directly proportional to $1/\varepsilon$ – but also directly proportional to the mole fraction of the organic solvent –; the stability of the complexes therefore increases with the percentage of alcohol (Tur'yan *et al.*, 1959–1960; Marple, 1969). In other cases, the relationship is linear only over a restricted range of composition of the mixtures of solvents.

On the other hand, it is not unusual for the part played by the electrostatic effect in the modification of the equilibrium constant to be substantially less than that due to the other effects of solvation, and for the overall modification to be in the opposite direction to that caused by the electrostatic effect. Thus, a complex such as $Hg(CN)_2$, which is very stable in aqueous solution ($K=10^{34.7}$ mol^{-2}l^2) becomes much less stable in ethanolamine ($\varepsilon=37.7$) and in ethylenediamine ($\varepsilon=14.2$), despite the low dielectric constant of these latter solvents in comparison with water: $K=10^{16.0}$ in ethanolamine, and $10^{11.6}$ in a mixture of equal weights of water and ethylenediamine ($\varepsilon=45$). We can attribute the greater part of the observed dissociating effect to the particularly strong solvation (by coordination) of the mercuric ion Hg^{2+} by the molecules of these organic nitrogenous solvents. Further, an investigation of the variation of $\log K$ in mixtures of water and ethanolamine or ethylenediamine, with progressive variation of the proportions, showed that there was a rapid decrease in stability on addition of a small proportion of organic solvent; the value of $\log K$ thus varies from 34.7 in pure water to 16.0 and 13.2 respectively for mixtures of only

10% by weight of ethanolamine or ethylenediamine. Stability passes through a minimum at about 50%, then increases again in mixtures which are more rich in the organic solvent; the variation of $\log K$ as a function of $1/\varepsilon$ is then almost linear, which appears to show that the electrostatic effect has become preponderant in this range of composition (Trémillon *et al.*, 1965).

Attempts have been made in order to palliate the insufficiency of the electrostatic theory to render an account of experimental variations in the equilibrium constants when the solvent is changed, by seeking to link these variations with other factors characteristic of the solvent, such as the dipole moment of its molecules, the activity of one of its constituents (in the case of mixtures of solvents), etc. More particularly, the influence of the structure of the solvent (associations by hydrogen bonds) has been considered and appears to be of great importance. For example, the formation constants $\log K$ of $Hg(SCN)_2$ and $Hg(SCN)_4^{2-}$ in mixtures of N-methylacetamide (NMA) and N,N-dimethylformamide (DMF) decrease as the percentage of NMA ($\varepsilon = 165$ at 40 °C) increases with respect to that of DMF ($\varepsilon = 35$ at 40 °C), from 0 to about 70% (the variation is approximately linear with $1/\varepsilon$ up to about 50% NMA). But in mixtures which are rich in NMA, $\log K_4$ and $\log K_2$ vary in the opposite direction, increasing with the percentage of NMA. This effect has been interpreted in the following manner: the degree of association of NMA molecules by hydrogen bonds increases with the percentage of NMA (DMF alone does not form hydrogen bonds). The SCN^- anions, principally solvated by hydrogen bonds, therefore undergo a correlative decrease in their solvation ($\log \gamma_t(SCN^-)$ increases); to the extent that solvation of other species does not alter greatly, the result is therefore an increase in stability of the complexes formed by that anion (Pucci, Vedel and Trémillon, 1969). A theory has recently been suggested which deals with this effect in a more quantitative manner (Vedel and Postel, 1970).

Despite these different approaches, it still appears impossible at the present time to give a general analytic theory providing a quantitative account of the modifications to equilibrium constants on change of solvent, in terms of the characteristic features of these solvents, dielectric constant in particular. More often, it is only possible to make a qualitative analysis to take into consideration all the effects observed, and predictions can only be made in a fairly rough manner.

Survey of the thermodynamic treatment of solutions in molten salts

We can consider an extension to solutions in ionized solvents, especially in molten salts, of the preceding thermodynamic treatment developed for solutions in molecular solvents. The high utilization temperatures of these media prevent the use of water as reference solvent for quantitative comparisons, which have meaning only at constant temperature.

Nevertheless, thermodynamic theories of solutions in molten salts exhibit some features specific of completely ionized media. We will confine ourselves to an outline survey of these features, giving an idea of the help which this can give towards the aim set out in this chapter.

In a general way, we find again in this treatment, as in those which have just been developed, the regard for connecting the physico-chemical behaviour of the solutes with their physical properties and in particular with the electrostatic interactions between ions.

Temkin's ideal model (1945)

Let us consider a molten salt which is completely ionized and is constituted of two ions A^+ and B^-, in which there is dissolved another similar molten salt constituted of the two ions X^+ and Y^-. In accordance with the quasi-crystalline structural model mentioned in the first chapter (Section 1.4), it can be said that, if the two cations and the two anions respectively have the same physical characteristics, the X^+ cations will be distributed completely at random over the sites of the cationic sub-lattice of the A^+, whilst the Y^- anions will for their part be distributed also completely at random over the sites of the anionic sub-lattice of the B^- ions. Under these conditions, it can be shown – by calculation of the total entropy of mixture in accordance with the number of possible distinct and equivalent configurations – that formation of a solution causes no energy effect; it can then be said that the effect of solvation is zero. The solution exhibits ideal behaviour whatever the concentration of the solute salt, and the chemical potential μ_{XY} of this latter depends only on the concentrations C_X of the cation and C_Y of the anion:

$$\mu_{XY} = \mu_{XY}^0 + 2.3RT \log C_X C_Y . \tag{V-40}$$

If in this case we take the pure molten salt XY as standard state, the concentrations are replaced by the mole fractions N_X and N_Y and the activity a_{XY} of the salt in solution is connected with the product of the mole fractions of the cation X^+ and of the anion Y^- by the simple equation:

$$a_{XY} = N_X N_Y . \tag{V-41}$$

It is to be noted that the mole fraction of the cation is defined with regard to all the cations only (since X^+ is mixed only with cations) and the mole fraction of the anion with regard to all the anions only. Thus, for a solution of X^+B^- in A^+ B^-, we have $N_B=1$, whence $a_{XB}=N_X$; for a solution of 1 mole of XY in 9 moles of AB, $N_X=0.1$ and $N_Y=0.1$, whence $a_{XY}=10^{-2}$.

In the case of a dissolved ion with a higher charge than the corresponding ion of the solvent, vacancies in its sub-lattice (cf. section I.4) are counted in the calculation of the mole fraction as if they were ions. Thus for a solution of salt XB_2 in AB, each X^{2+} ion creates a vacancy; if the total numbers of moles of AB and XB_2 are respectively n_{AB} and n_{XB_2}, the total number of vacancies is equal to the number of cations X^{2+}, say n_{XB_2}, whence:

$$N_X = \frac{2n_{XB_2}}{n_{AB} + 2n_{XB_2}}. \tag{V-42}$$

It is therefore the equivalent ion fraction which should be considered, rather than the mole fraction (Førland, 1957).

The choice of the pure substance as standard, rather than a concentration of 1 mol l^{-1} or 1 mol kg^{-1} – only the value of μ^0_{XY} would be different – has the interesting feature of providing a reference state independent of the solvent.

It has been found by experiment that practically ideal behaviour in accordance with this model takes place only in a few binary mixtures: such as AgCl+LiCl, $CdCl_2$+ $PbCl_2$, $CdCl_2$+AgCl, $PbCl_2$+LiCl, etc.

Non-ideal solution in a molten salt.
Activity coefficient

For most actual solutions, divergences from the ideal behaviour postulated by Temkin have been observed. Solution formation is accompanied by an energy effect to which there corresponds an 'excess chemical potential' of the solute, which can be interpreted as being the result of solvation effects.

Contrary to the case of electrolytic solutions (molecular solvents), we can introduce into the medium one ionic species only. For example, when dissolving a salt X^+B^- in a solvent A^+B^-, the X^+ cation is the only solute introduced.

We can assume here again that behaviour is ideal in a sufficiently dilute solution and then regard the activity of X^+ as equal to its concentration. Under these conditions, the reference state is that of X^+ in an infinitely dilute solution in AB. Due to solvation effects, the chemical potential μ_X^0 of X^+ in the standard state obtained by extrapolation, in this reference state, up to ion fraction of unity is different from the standard chemical potential μ_X^* of X^+ in pure XB. The difference $\mu_X^0 - \mu_X^* = \mu_X^E$ is the excess chemical potential of the solute; it corresponds to the free enthalpy of transfer of X^+ from its pure (liquid) salt to an infinitely diluted solution in the molten salt AB.

The relative activity of X^+ in solution can be defined in two ways, one with regard to the standard state μ_X^0, the other with regard to the standard state μ_X^*. In the first case:

$$\mu_X = \mu_X^0 + 2.3RT \log a_X \qquad \text{(V-43)}$$

with, in an infinitely diluted solution, $a_X \to N_X$. In the second case:

$$\mu_X = \mu_X^* + 2.3RT \log a'_X \qquad \text{(V-44)}$$

(where $a'_X = 1$ in the pure substance XB). We can then express μ_X^E in the form of an activity coefficient (ratio activity/concentration), which we will denote by γ_X:

$$\mu_X^E = 2.3RT \log \frac{a'_X}{N_X} = 2.3RT \log \gamma_X . \qquad \text{(V-45)}$$

In Temkin's ideal model, $\mu_X^E = 1$, and hence $\gamma_X = 1$. Solvation can be regarded as positive when $a'_X < N_X$ (the X^+

activity is less than when it is the only cation); then $\gamma_X < 1$.

For example, for the systems quoted in the previous paragraph as behaving almost ideally in the sense of the Temkin model, the activity coefficient of one of the constituents diluted in the other remains between 0.9 and 1. But for a dilute solution of $CdCl_2$ in molten KCl, an activity coefficient γ_{Cd} of the order of 0.2 has been observed at 900 °C (Barton and Bloom, 1959), showing that there is a fairly strong solvation of Cd^{2+} by the Cl^- ions (in the form of complex species $Cd\,Cl_4^{2-}$ or $Cd\,Cl_6^{4-}$).

The activity coefficient γ could be called the 'transfer activity coefficient' as in the case of molecular solvents. The only difference is that the reference state in this case is the pure substance and no longer an infinitely diluted aqueous solution.

Electrostatic contribution to the activity coefficient of an ion dissolved in a molten salt

In order to give at least a partial explanation of the excess chemical potential, we can introduce modifications to the various types of interaction between the ions, and in particular the electrostatic coulombic forces, which are the main factors in the simplest cases.

When we introduce the salt X^+B^- into the solvent A^+B^-, the environment of the solute cation X^+ is modified, at least as far as the closest cations are concerned (since the anions are all identical); in place of other neighbouring X^+ cations as in pure XB, the dissolved X^+ is surrounded by the A^+ cations of the solvent. This leads to changes in the coulombic energies of electrostatic repulsion between these cations; Førland (1955–1957) and later Blander (1961) have sought to estimate these so as to account for the excess chemical potential in the case where the solute ion is subjected only to coulombic interactions.

Førland evaluated the change in Coulombic energy which corresponds to the exchange process between triplets of ions (treated as rigid spheres in contact):

$$\tfrac{1}{2}X^+B^-X^+ + \tfrac{1}{2}A^+B^-A^+ \rightarrow X^+B^-A^+ \qquad \text{(V-46)}$$

and multiplies this energy by the number of cations A^+ which are the closest neighbours of X^+. We thus reach a final expression for the excess chemical potential:

$$\mu_{el}^{E} \approx -k\left(\frac{1}{d_{AB}} + \frac{1}{d_{XB}}\right)\left(\frac{d_{AB} - d_{XB}}{d_{AB} + d_{XB}}\right)^2 \qquad \text{(V-47)}$$

where d_{AB} is the sum $r_A + r_B$ of the ionic radii of the ions A^+ and B^-, and d_{XB} is the sum $r_X + r_B$. The factor k is a constant dependent on the geometrical structure of the ions around the solute (it is connected with the factor of $1/d$ in the expression for the reticular energy of the crystal).

Blander has suggested completion of the calculation, in the case of an extremely dilute solution of X^+ ions, by allowing for the coulombic interactions at greater distances (by carrying out the calculation for an infinite linear chain of adjacent spherical ions... $A^+B^-A^+B^-X^+B^-$ $A^+B^-A^+$...).

The expression (V-47) given by Førland, like Blander's calculation, leads to several interesting conclusions:

1. When the X^+ ion has the same dimensions as the A^+ ion of the solvent, $\mu_{el}^E=0$, and we still have the Temkin ideal solution.

2. The divergence from the ideal state becomes more pronounced the greater the difference in dimensions between the solute ion and the solvent ion of the same sign.

3. For given cations X^+ of solute and A^+ of solvent, the divergence from the ideal state become larger as the counter-ion B^- becomes smaller.

4. The value of μ_{el}^E is always negative (therefore $\gamma_{el}<1$). The differences in electric charge of the ions also naturally plays a part (in the value of k).

But, as in the case of ion solvation in a molecular solvent the coulombic forces are usually not the only ones operating. On one hand, the anions can often be polarized by the surrounding cations, and polarization forces then have to be added to the above; an approximate expression (Lumsden, 1961) for the energy which they develop leads nevertheless to conclusions which are similar to the above, the only difference being that the size effect of the common anion is greater. On the other hand, other types of force must also be added: especially coordinating bonds (complex formation) which can be very much the major effect in many cases.

In the present state of knowledge, there is no adequate theory accounting completely for the effects of solvation, no more in molten salts than in molecular solvents, and in this area of thermodynamics empiricism still plays an important part.

Experimental determination of activity coefficients of ions dissolved in molten salts

As in the case of molecular solvents, it is shown that the only experimentally accessible activity coefficients are those of electrically neutral solutes, hence mean activity coefficients in the case of ionized solutes. Let us consider what is in fact the most suitable method for this determination: electromotive force measurement of cells.

For example, let MX_n be a metallic halide dissolved in a molten alkali halide A^+X^-. The electromotive force E of the cell:

$$\mathrm{M(sol\ or\ liq)} \mid \mathrm{MX}_n \text{ diluted in } \mathrm{A^+X^-(liq)} \mid \mathrm{X_2}\ \text{(gas, 1 atm); C (graphite)}$$

varies with the solute activity, that is the solvated metallic ion activity; the other activities are equal to unity. This electromotive force can be expressed in two ways, depending on the standard state used for M^{n+}:

$$E = E_M^0 - \frac{2.3RT}{nF} \log a_M \qquad \text{(V-48)}$$

or

$$E = E_M^* - \frac{2.3RT}{nF} \log a'_M . \qquad \text{(V-49)}$$

In the first case, since the activity a_M is defined with regard to an infinitely dilute solution of M^{n+} in molten A^+X^-, E_M^0 is the value of the electromotive force obtained by extrapolation of E in very dilute solution (ideal behaviour $a_M = N_M$) to the mole fraction $N_M = 1$. In the second case, since the activity a'_M is defined with regard to the pure substance MX_n, E_M^* is the electromotive force of the cell:

M(sol or liq) | pure $MX_n(a'_M = 1)$ ||
|| A^+X^-(liq) | $X_2(g, 1\text{ atm})$; C

after deduction of the liquid junction potential between the two half cells. Unfortunately, this latter cannot be determined and prevents our obtaining an exact value of E_M^*, from which it might have been possible to deduce the activity coefficient γ_M of the metallic ion:

$$\log \gamma_M = \frac{\mu_M^E}{2.3RT} = \frac{nF}{2.3RT} (E_M^* - E_M^0) . \qquad \text{(V-50)}$$

If we consider, in place of this, the electromotive force $E_{MX_n}^*$ of the cell:

$M(s$ or liq) | pure MX_n | $X_2(g, 1\text{ atm})$; C

(a cell without liquid junction), its value depends not only on the state of the metallic ion but also on that of the halide anion, since this latter is not under the same activity conditions when in pure MX_n as when in the molten alkali halide. Having regard to this fact, we can express the electromotive force E in the form:

$$E = E^*_{MX_n} - \frac{2.3RT}{nF} \log a'_{MX_n} = E^0_M - \frac{2.3RT}{nF} \log a_{MX_n}. \tag{V-51}$$

where E_{MX_n} is the electromotive force corresponding to unit activity of the salt MX_n,

Since the difference $E^*_{MX_n}$—E^0_M corresponds to the excess chemical potential $\mu^E_{MX_n}$ of the solute as a whole, and therefore to the activity coefficient γ_{MX_n}:

$$\log \gamma_{MX_n} = \frac{\mu^E_{MX_n}}{2.3RT} = \frac{nF}{2.3RT} (E^*_{MX_n} - E^0_M). \tag{V-52}$$

We note that the value of $E^*_{MX_n}$ can be determined without recourse to experimental measurements, if we know the standard free enthalpy $\Delta G^*_{MX_n}$ of formation of the halide MX_n (in the pure state) starting from the metal M and the halogen X_2 (in the physical conditions of the cell under consideration). It is easily shown that

$$E^*_{MX_n} = - \frac{\Delta G^*_{MX_n}}{nF}. \tag{V-53}$$

In order to determine the value of the activity coefficient γ_M of the ion considered individually, similar methods to those set for the case of molecular solvents must be considered, if possible (reference couple ferrocene-ferricinium$^+$ for example, if the temperature is not too high).

V.2. Correlation Between Acid-Base Properties in Different Solvents

In order to understand the reasons for which acid-base properties undergo substantial changes when the solvent is changed, and if possible to be in a position to interpret and predict these changes, it is necessary to determine a correspondence between these properties in a non-aqueous solvent and those in aqueous solution (for example), taken by convention as a reference.

The fundamental correspondence to be settled is that between the pH scales (scales of relative activity of the hydrogen ion) in the various solvents. It proceeds as a consequence of the general thermodynamic considerations which have just been described in connection with ion solvation in molecular solvents.

Relationship between the pH scales in different solvents

Having regard to the relation between chemical potential and activity of a species, we can give a 'thermodynamic' definition of the pH in a solvent*:

$$\mathrm{pH} = -\log a_{\mathrm{H}^+} = \frac{\mu^0_{\mathrm{H}^+} - \mu_{\mathrm{H}^+}}{2.3RT}. \qquad \text{(V-54)}$$

A pH scale therefore corresponds to a scale of chemical potential for the hydrogen ion, one unit of pH corresponding to one unit of $-\mu/2.3RT$. The origin of the scale, pH=0, corresponds to the standard state $\mu^0_{\mathrm{H}^+}$ in the solvent under consideration.

As for all species, the standard chemical potential of H^+ varies with the solvent, being lower as solvation of H^+ has more energy, that is as the solvent molecules are more strongly basic. Thus, a solution of a strong acid in a solvent S at such a concentration that pH(S)=0 ('ideal' solution of HS^+ 1 M) and an aqueous solution of a strong acid at such a concentration that pH(aq)=0 ('ideal' solution of H_3O^+ 1 M) do not have the same absolute activity of H^+.

Qualitatively, the difference arises from the fact that the acidity of the aqueous solution corresponds to the acid-base couple H_3O^+/H_2O, whilst that of the non-aqueous solution corresponds to a different acid-base couple, HS^+/S. If the solvent S is less basic than water, the absolute activity of H^+ in the non-aqueous solution is

* This enables us to state that pH, a thermodynamic quantity (therefore an energy quantity), retains a meaning even under conditions where the concentration of free solvated H^+ can no longer be physically defined ($\mathrm{pH} \gg 24$, that is 'less' than one true solvated H^+ ion in the volume of solution used).

stronger than that in the aqueous solution; the opposite is the case if S is more basic than water. The first class includes nitriles, ketones, and especially carboxylic, sulphuric and hydrofluoric acids, etc; the second class includes liquid ammonia, amines, pyridine, HMPA. Alcohols and amides have basic strengths not very much different from that of water.

The scale of chemical potential of the hydrogen ion (whatever the solvating medium) can be regarded as an absolute scale of acidity, and on this scale we can show the values – assuming that they can be determined – of the standard chemical potential of H^+ in aqueous solution, $(\mu^0_{H^+})_{water}$ (to which corresponds pH(aq)=0), or in another protophilic solvent S, $(\mu^0_{H^+})_S$ (to which corresponds pH(S)=0). According to the relationships of Section V.1, the difference in chemical potential $(\mu^0_{H^+})_S - (\mu^0_{H^+})_{water}$ is equal to the free enthalpy of transfer of H from one solution to the other. By reference to the definition of the transfer activity coefficient $\gamma_t(H^+)$, we can conclude that the non-aqueous solution of pH(S)=0 has the same absolute acidity (that is the same chemical potential of H^+) as an aqueous solution of pH(aq)$=-\log\gamma_t(H^+)$.* Thus, the highest acidity that can be attained in the solvent – by means of concentrated strong acids – corresponds to pH(aq) near to $-\log\gamma_t(H^+)$; this is a medium of absolute acidity which is greater or less than that of aqueous strong

* A hypothetical solution if $\gamma_t(H^+) \gg 1$, that is if the solvent S is much less basic than water.

acid solutions, depending on whether the solvent S under consideration is less basic or more basic than water [$\log \gamma_t(H^+) > 0$ or < 0].

Further, in the case of an amphiprotic solvent (HS), the highest basicity (the weakest acidity) that can be attained – by means of concentrated solutions of strong base S^- – corresponds to pH(aq) close to $pK_S - \log \gamma_t(H^+)$. In comparison with the most basic media available in aqueous solution – concentrated solutions of OH^- ions, pH(aq)$\approx$ 14 – we have a medium with a basicity greater or less than that of water, depending on whether the solvent HS is a weaker or a stronger acid than water.

Lastly, all the values of pH(S) available in the solvent, within the limitations caused by the acid-base properties of its molecules, are connected with the values of pH(aq) by the equation:

$$\mathrm{pH(aq)} = \mathrm{pH}(S) - \log \gamma_t(H^+). \qquad \text{(V-55)}$$

Modification of the properties of acid-base couples on change of solvent

After relating (theoretically) the values of pH, we will now seek to compare the absolute acidities ('acidity levels') of solutions composed of the same mixture of an acid and its conjugate base in a fixed proportion – it will possibly be a 'buffer' of pH – in water on the one hand and in another solvent S on the other hand.

1. Let us first consider the case of a couple where the acid

and the base, both weak in water, remain so in the solvent S under consideration (that is they undergo only solvation and negligible solvolysis).

The values of the acidity constant K_A, characteristic of the couple in each of the two media, are connected by:

$$K_A(S) = K_A(\text{aq}) \frac{\gamma_t(\text{HA})}{\gamma_t(\text{H}^+)\cdot\gamma_t(\text{A})} \tag{V-56}$$

(where HA and A are respectively the acid and its conjugate base, without specifying their ionic charge). Let us suppose that the proportion HA/A is such that in aqueous solution $\text{pH}=\text{p}K_A(\text{aq})$; in the non-aqueous solution, we shall have in the same way $\text{pH}(S)=\text{p}K_A(S)$. To this latter value there corresponds a value of pH(aq) defined by:

$$\text{pH}(\text{aq}) = \text{pH}(S) - \log\gamma_t(\text{H}^+) = \text{p}K_A(S) + \\ - \log\gamma_t(\text{H}^+). \tag{V-57}$$

Using Equation (V-56), we have:

$$\text{pH}(\text{aq}) = \text{p}K_A(\text{aq}) + \log\frac{\gamma_t(\text{A})}{\gamma_t(\text{HA})}. \tag{V-58}$$

The change in the absolute acidity when we pass from aqueous solution to solvent S is thus given by:

$$\Delta(\text{pH}) = -\frac{\Delta\mu_{\text{H}^+}}{2.3RT} = \log\gamma_t(\text{A}) - \log\gamma_t(\text{HA}). \tag{V-59}$$

We therefore obtain a medium which is more acidic

than in water if $\gamma_t(\mathrm{A})<\gamma_t(\mathrm{HA})$, and less acidic in the opposite case. If $\gamma_t(\mathrm{A})=\gamma_t(\mathrm{HA})$, that is if solvation of the acidic form and of the basic form vary in the same way, the absolute acidity remains unaltered, and the acid-base couple in question can serve as an absolute reference of acidity for experimental correlation between the pH scales (see later).

Amongst the factors capable of causing this change in acidity, the difference in dielectric constants is the one most frequently considered. We can allow for its influence, due to the laws previously seen (cf. Section V.1), which state the contribution of the electrostatic effect to the values of the transfer activity coefficients $\gamma_t(\mathrm{A})$ and $\gamma_t(\mathrm{HA})$. This influence differs according to whether the acid-base couple in question belongs to one or other of the following types:

(a) Couples consisting of a cationic acid and a molecular base (HA^+/A): having regard to the effect of variations of ε on the value of the only transfer activity coefficient $\gamma_t(\mathrm{HA}^+)$, expressed by equation (V-13), the change in dielectric constant, from $\varepsilon_{\mathrm{water}}$ in aqueous solution to ε_S in solvent S, leads to a change in acidity given by:

$$\Delta(\mathrm{pH})_{\mathrm{el}} = -\frac{k}{r_{\mathrm{HA}^+}}\left(\frac{1}{\varepsilon_S} - \frac{1}{\varepsilon_{\mathrm{water}}}\right). \qquad \text{(V-60)}$$

There is an increase in acidity if the dielectric constant of the solvent S is less than that of water.

(b) Couples consisting of a molecular acid and an anionic base (HA/A^-): in this case, changes in ε affect the value of $\gamma_t(A^-)$, and the change in acidity is given by:

$$\Delta(\mathrm{pH})_{\mathrm{el}} = \frac{k}{r_{\mathrm{A}^-}}\left(\frac{1}{\varepsilon_S} - \frac{1}{\varepsilon_{\mathrm{water}}}\right) \qquad \text{(V-61)}$$

and corresponds on the contrary to a fall in acidity when the dielectric constant of S is less than that of water.

(c) We can also be concerned with couples of type HA^{2+}/A^+ or HA^-/A^{2-}. In this latter case, for example, having regard to the double effect of changes in ε on both $\gamma_t(A^{2-})$ and on $\gamma_t(HA^-)$, we have:

$$\Delta(\mathrm{pH})_{\mathrm{el}} = k\left(\frac{4}{r_{\mathrm{A}^{2-}}} - \frac{1}{r_{\mathrm{HA}^-}}\right)\left(\frac{1}{\varepsilon_S} - \frac{1}{\varepsilon_{\mathrm{water}}}\right). \qquad \text{(V-62)}$$

Notes. It must not be forgotten that the change in dielectric constant is also responsible for part of the value of $\gamma_t(H^+)$, that is, for part of the difference between pH(S) and pH(aq).

In this way, for a couple of type HA^+/A, the value of the acidity constant pK_A, corresponding to an 'iso-ionic' equilibrium, is practically independent of variations of ε, although the absolute acidity level of the mixtures HA^+/A is affected by these variations.

It must also not be forgotten than changes in the dielectric constant are only one of the factors which modify the acidity, and that this factor would not suffice to account for the effects observed by experiment. This will be seen from later examples.

2. *Levelling effect of the solvent.* The connection between the absolute acidity levels of solutions of a given acid-base couple in water and in another solvent S, as given by Equation (V-59), is valid only if the solvolysis effect on the acid and on the base remain small in the two media. In the opposite case, what is known as the 'levelling' effect of the solvent intervenes, enabling us to understand why an acid (or base) which is strong in water can become weak in another solvent, or vice versa.

Let us consider for example a couple HA/A, where the acidic form and the basic form are both weak in aqueous solution. By considering only the effect of changes in solvation given by equation (V-59), there should be a certain level of absolute acidity corresponding to a mixture HA+A in another solvent. But:

(a) If this level is more acidic than the most acidic level attainable in solvent S – defined by $-\log \gamma_t(H^+)$ –, it will not be possible to observe it in practice, and from being weak in water, the acid HA will become strong in S. The acidity is no longer fixed by the couple HA/A, but by the couple HS^+/S; it is the result of the solvolysis, substantial if not quantitative, of HA by S molecules.

All acids of this type, which by virtue of their different values of pK_A(aq) give different values of acidity in water, give the same acidity level in solvent S (at the same concentration), corresponding to the couple HS^+/S (the acidity level of all strong acids in S). The acidity of all these acids is thus 'levelled' by the basicity effect of solvent S.

(b) In a similar manner, if the hypothetical level of absolute acidity of the mixture HA+A when transferred from water to an amphiprotic solvent HS is more basic than the most basic level possible in this solvent – defined by $pK_S - \log \gamma_t(H^+)$ –, it will not be observable, and from being weak in water, the base A will become strong in HS. This time, the basicity is fixed by the couple HS/S^-, due to solvolysis of A by HS molecules.

All bases of this type, with different acidity levels in water, give the same acidity level in HS (at the same concentration), corresponding to the couple HS/S^- (the acidity level of all strong bases in HS). The basicity of all these bases is thus levelled by the acidity effect of the solvent HS.

On the contrary, a couple which is 'levelled' in aqueous solution (strong acid or strong base) can become, in another solvent, – from the same reasoning, interchanging the parts played by water and by the other solvent – a couple weak acid/weak base, and two acids which are strong in water and in consequence lead to the same acidity, can give different acidity levels in another solvent.

EXPERIMENTAL REALIZATION OF THE CORRELATION AND PRACTICAL CONSEQUENCES.

Experimental methods used for comparison of the absolute acidity levels in different solvents

1. In order to carry out this comparison, it is first necessary to determine experimentally the values of the transfer activity coefficient $\gamma_t(H^+)$ of the hydrogen ion in these media. The problems involved in this determination and the methods used in general to solve them have previously been mentioned. The general method, which seems at present to be the most suitable in principle, is the determination of $R_0^0(H^+)$ – by measurements of electromotive forces of cells including a normal hydrogen electrode in the solvent concerned (when it operates suitably, which is not the case for all solvents) and the reference (of potential) ferrocene-ferricinium$^+$ electrode or any other analogue – assuming that $-\log\gamma_t(H^+) \approx R_0^0(H^+)$.

If the hydrogen electrode enables us to obtain the value of $R_0^0(H^+)$ in the solvent in question (with regard to water) it also enables us to measure the values of pH(S) and in consequence to connect these values with the scale of pH(aq), that is to compare any acidity level in the non-aqueous solution with the acidity level in aqueous solution.

2. Another method, older, of experimental comparison of absolute acidity levels in different solvents is given by determination of the acidity function H_0 defined by Ham-

mett (1932–1935). This determination calls for spectrophotometric (absorptiometric) determinations, which are easy to carry out in most solvents, by means of pH indicators of neighbouring chemical compositions – in particular, there is a series of substituted nitroanilines, of which the basic molecular forms and the conjugate protonated (cationic) forms have different absorption spectra, and the pK_A values in various solvents have been found to retain approximately constant differences.

By denoting, for one of these indicators, the two forms, base and conjugate acid, by I and HI^+ respectively, and the value of pK_A of the couple HI^+/I in aqueous solution by $pK_I(aq)$, Hammett defined the acidity function:

$$H_0 = pK_I(aq) + \log \frac{[I]}{[HI^+]}. \qquad (V\text{-}63)$$

The value of H_0 corresponding to a medium of given acidity, whatever the solvent, is obtained by absorptiometric measurement of the ratio $[I]/[HI^+]$. Measurements are significant only if this ratio is not greatly different from unity, that is for values of H_0 near to $pK_I(aq)$ (± 1 approximately). By means of a series of indicators with maximum differences of 1 to 2 units between two successive values (see Table X), it is therefore possible to cover a relatively extended range of values of H_0 (cf. Paul and Long, 1957).

The quantity H_0 may be regarded as equal to the pH in dilute aqueous solution (insofar as the activity coefficients

TABLE X

Some molecular bases (I) used as indicators for determination of the acidity function H_0

I	pK_I(aq)
p-aminoazobenzene	2.76
benzeneazodiphenylamine	1.42
p-nitroaniline	0.99
o-nitroaniline	—0.29
4-chloro-2-nitroaniline	—1.03
p-nitrodiphenylamine	—2.50
2,4-dichloro-6-nitroaniline	—3.32
2,4-dinitroaniline	—4.53
2,4,6-trinitro-N,N-dimethylaniline	—4.81
p-benzoyldiphenyl-6-bromo-2,4-dinitro aniline	—6.71
2,4,6-trinitro aniline	—9.41

From Paul and Long, 1957.

of I and of HI^+ remain close to unity). In non-aqueous solutions, if we assume that the value of the ratio $[I]/HI^+]$ corresponds to a level of absolute acidity which is invariable whatever the solvent, the values of H_0 will remain equal to those of pH(aq) corresponding to pH(S) which can be measured by another way. In particular, non-aqueous solutions where pH(S)=0 will each be characterized by a value H_0 which, to the extent that the preceding hypothesis is valid, represents $-\log \gamma_t(H^+)$ in the medium under consideration.

The hypothesis which permits us to regard $-\log \gamma_t(H^+)$ as equivalent to H_0^0 is of the same nature as that which allows us to regard it as equivalent to $R_0^0(H^+)$. The assumption of a level of absolute acidity independent of the solvent, for a mixture I/HI^+ of given composition, amounts to – as when we assume an invariable potential for a given ferrocene-ferricinium$^+$ mixture – assuming that the ratio of the transfer activity coefficients of the two conjugate species I and HI^+ is invariable (cf. Equation V-59). We must, however, expect this hypothesis to be less well verified in the case of pH indicators than in that of the ferrocene-ferricinium$^+$ couple, and it is probable that the values of $R_0^0(H^+)$ are closer to $-\log \gamma_t(H^+)$ than those of H_0^0. This statement is based on consideration of structural differences, which must result in solvation effects which are much more pronounced for the indicators than for ferrocene and an influence of change of dielectric constant affecting HI^+ cations more than it affects ferricinium$^+$.

Experimental measurements have in fact shown fairly substantial differences, as large as one or two units, between the values of H_0^0 and those of $R_0^0(H^+)$ determined in several binary mixtures of solvents, such as water+sulphuric acid, water+ethanol, water+acetone, water+ +acetonitrile (Vedel, 1967). The very approximate nature of the experimental correlation between the pH scales in different solvents is a consequence of these differences.

Apart from its probably better theoretical basis as has just been discussed, the correlation method based on the ferrocene-ferricinium$^+$ couple has other advantages over the acidity function H_0. The first method requires only one potentiometric determination, and not the successive approximations of the indicator method, when we have to link aqueous solutions with a medium of acidity very different from those attainable in water. There is also a limitation on the use of the H_0 function in the direction of the most basic media (greater than pH(aq)$\approx$9–10), in which no molecular indicator can any longer exist in the protonated state HI^+. Indicator couples of types I^-/HI and even I^{2-}/HI^-, to which there correspond acidity functions H_- and H_{2-}, are then used, but in these cases the electrostatic interactions between solute and solvent are different (frequently much larger) and these acidity functions probably diverge from the values of pH(aq) even more than H_0 does. Nevertheless, as only a few determinations of $R_0^0(H^+)$ have been made up to now for pure solvents, the values of the acidity functions H, of which many more are available, have been found extremely useful in determining at least the order of magnitude of the absolute acidity differences between two media.

Ranges of acidity available in the usual solvents

Experimental results, obtained by one or other of the methods which have just been described, enable us to place approximately the ranges of acidity available in the different non-aqueous solvents, on a scale of absolute acidity (that is of chemical potential of H^+) graduated in values of pH(aq).* Each of these ranges extends over about pK_S units (for amphiprotic solvents), between the lowest values corresponding to the most concentrated solutions of strong acid in the solvent under consideration, and the highest values corresponding to the most concentrated solutions of strong base. If the solvent is not protogenic, the range is in principle unlimited in the direction of the most strongly basic media; it is limited in practice by decomposition of the solvent, or simply because of the impossibility of obtaining stronger dissolved bases.

Figure 13 shows the results corresponding to some of the most common molecular solvents. The correlation principle is shown as a diagram at the top of the figure. The range of acidity available in aqueous solution is limited, projected onto all the ranges of the other solvents, by the two discontinuous lines perpendicular to the axis of chemical potential; the media with absolute activity greater than that of aqueous solutions of strong acid are to the left of the left-hand line; media with absolute basicity greater than that of aqueous solutions of strong base are to the right of the right-hand line.

It is thus seen that certain solvents enable us to produce solutions which are much more acidic than aqueous solutions of strong acid; this is the case with formic, acetic, sulphuric and hydrofluoric acids, DMSO, and nitrometh-

* Graduation in values of pH(aq) appears the most convenient and the most informative for comparison of non-aqueous media with traditional aqueous media.

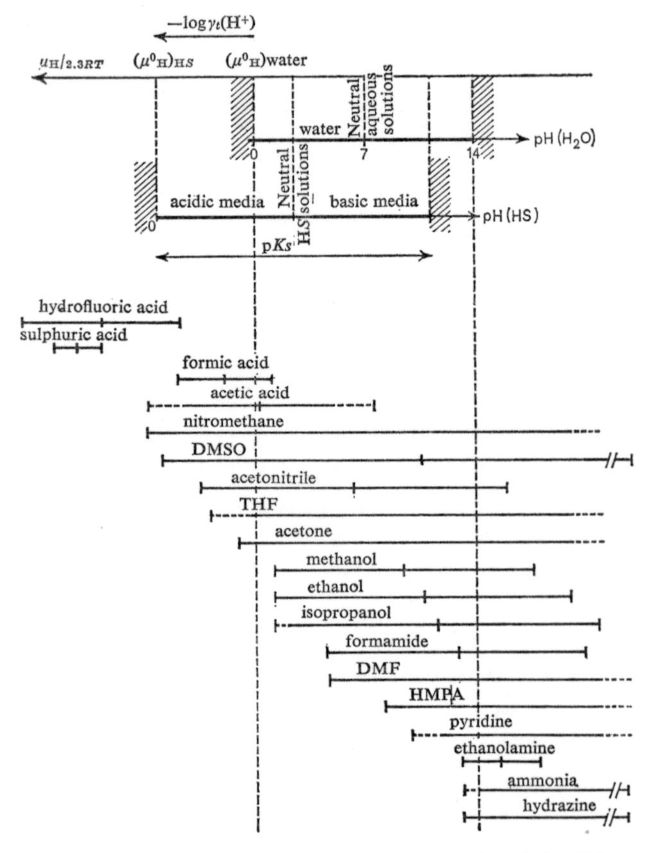

Fig. 13. Correspondence between the scales of pH in different molecular solvents, at ordinary temperature.

ane. These solutions are 'superacid' media, characterized by large negative values of pH(aq) (up to −15 in hydrofluoric acid, and −13 in sulphuric acid). It is evident that these solvents are useful for chemical processes where a high protonic activity is necessary.

On the other hand, other solvents enable us to produce solutions which are much more basic than aqueous solutions of strong base; this is the case with ammonia and the amines, and with non-protogenic solvents: acetone, DMF, THF, HMPA, pyridine, etc. These solutions are 'superbasic' media, characterized by very high values of pH(aq), the maximum reached being of the order of approximately 40.

We can quote some experimental values $R_0^0(H^+)$ which were used in construction of the diagram in Figure 13: acetic acid −7.0 (Durand, 1970) (but the substantial formation of ion-pairs in this solvent prevents us from reaching values lover than $pH(aq) \approx -5$); DMSO −5.8 (Le Demezet, 1970); acetonitrile −3.7 (Strehlow *et al.*, 1960); THF −3.0 (Perichon, 1970); NMA 5.5 and DMF 4.6 (Pucci, 1968), HMPA 8.5 (Madic and Trémillon, 1968); hydrazine 13.2 (Izmailov, 1963).

A similar correlation made for mixtures of solvents in variable proportions leads to graphs such as those shown on Figure 14, for water+ethanol and water+acetone mixtures. The progressive change in the range of absolute acidity as we pass from pure water to pure organic solvent are shown. Absolute acidity levels corresponding to

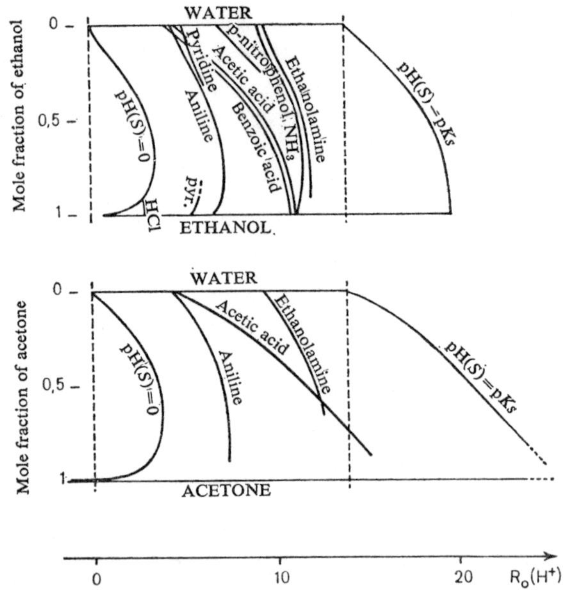

Fig. 14. Correlation between acid-base properties in water + ethanol and water + acetone mixtures.

pH(S)=0 have been located on the graphs of Figure 14 by adoption of the experimental variations of $R_0^0(\mathrm{H}^+)$ shown on Figure 12; the values of pK_S are those of Table IV. Some variations of acidity level of buffer mixtures have been shown by graphing, starting at pH(S)=0, the values of pK_A of the corresponding acid-base couples (given in Table V). We note in particular the relative variation in acidity level of couples of type HA/A$^-$ in comparison with the variations with couples of type HA^+/A; it can be at least partly attributed to the effect of a fall in dielectric constant when the proportion of organic solvent increases.

Consequences for application to acidimetric titrations

The levelling effect of a solvent, and of water in particular, has a harmful effect on the main application of acid-base reactions, acidimetric titrations: this consists of preventing differentiation between strong acids or between strong bases, which behave all in the same way (either as the protonic form of the solvent for strong acids or as the base conjugate to the solvent for strong bases). In addition, the limitation of the acidity scale on the side of the most acidic media does not permit sufficient neutralization by acidification of the weakest bases, and the limitation on the side of the most basic media does not permit neutralization of the weakest acids, and in consequence prevents carrying out of titrations in these cases.

These problems are solved by making use of the knowledge on the various non-aqueous media. Use of a solvent giving access to superacid media gives the possibility:

(a) of distinguishing between acids which are strong in aqueous solution, but which become weak, with different pK_A values, in this other solvent (acetic acid, for example; see Figure 5, which shows titration curves – variation of pH during neutralization – of five acids which are strong in water, and clearly distinguished in acetic acid).

(b) of neutralizing and hence titrating (titration curve with large change in pH in the neighbourhood of the equivalence point) bases which are too weak in water (for example: aniline and its substitution products, pyridine, quinoline, urea, etc., which can be titrated by $HClO_4$ in acetic acid).

In the same way, use of a solvent giving access to superbasic media gives the possibility of differentiating between bases which are strong in aqueous solution, and of neutralizing acids which are too weak in water (phenols, alcohols and thiols, etc.). For example, urea and ethanol can be neutralized as acids by amide anion (strong base) in liquid ammonia.

An amphiprotic solvent, which causes a double levelling effect, can be used for acidimetric titrations only if its acidity range (measured by pK_S) is sufficiently extended. From this view-point, many solvents, and especially mixtures of solvents, have too small values of pK_S and cannot be used (sulphuric acid and its mixtures with water, mixtures of water+acetic acid or water+ethanolamine, etc). The solvents which seem in principle to be the most promising are those which are non-protogenic and only slightly protophilic.

Prediction of the possibilities offered by non-aqueous solvents for carrying out acid-base reactions, together with the resulting applications, does not, however, depend only on consideration of the acidity ranges (compared in extent and in position on the absolute acidity scale). Regard must also be paid to the solvent influence on the more specific properties of the acid-base couples considered. We know that unfortunately this is a topic which still avoids any complete calculated prediction. Only the effect of changes in dielectric constant can be approximately taken into account. This is already an appreciable step forward, and thus we can predict, for example, that a solvent of lower dielectric constant than water renders molecular acids apparently weaker than in water, in the opposite direction to cationic acids. Anionic acids are still more weakened, in such a way that, in the case of molecular diacids for example, the difference between their two pK_A values can be greatly increased (thus, sulphuric, succinic and phthalic acids have two clearly separated acidities in isopropanol, $\varepsilon=18$, and in acetone, $\varepsilon=20.7$).

But formulas expressing the pK_A variation with the reciprocal of the dielectric constant provide only in exceptional cases a representation and an explanation, even approximate, of the experimental values. We will show the importance of the other effects by means of the following example. At a temperature of 100 °C, water and molten acetamide have approximately the same dielectric constant: $\varepsilon \approx 59$ for acetamide and 55 for water. Any difference between the properties of acid-base couples in these two media cannot be due to an electrostatic effect. By using the experimental values $R_0^0(H^+)=4.7$ and $pK_S=14.6$ for acetamide, $pK_{H_2O}=12.3$ (at 100 °C), together with different pK_A values for acid-base couples, determined in the two solvents at 100 °C (for water by extrapolation from

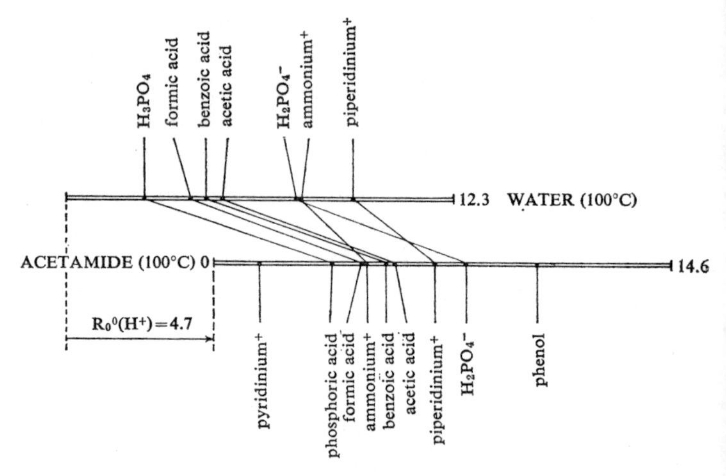

Fig. 15. Correlation between acid-base properties in water and in molten acetamide at 100°C. (From Guiot and Trémillon, 1968).

values at lower temperatures), we obtain the correspondence between the two acidity scales as shown on Figure 15. Each couple HA/A considered is shown at a level corresponding to pH(S). Thus, there are appreciable differences between the two media, which can be accounted for only by specific solvation effects (Guiot and Trémillon, 1968).

V.3. Correlation Between Other Properties

The method mentioned for quantitative comparison of

acid-base properties in different solvents can be extended to other types of property and in particular to the solvo-acidity phenomena described in Chapter III.

For each type of phenomenon concerned with exchange of an ionic species, the correlation is made by using, as in the case of the H^+ ion, the differences in standard chemical potential of the ionic species in question in each of the solvents considered. For example, for Cl^- exchange (chlorocomplex formation in general, and chloroacidity phenomena in the case of solvents donor of Cl^-), we will use a chemical potential scale of the Cl^- ion; the true pCl^- scales in each solvent will be placed so as to correspond, from the values of the free enthalpy of transfer of Cl^- from one solvent to another; each acceptor-donor couple of this ion will be placed on each pCl^- scale, at a position being determined by the characteristic equilibrium constant of the couple in the solvent in question; the difference of Cl^- chemical potential exhibited by the couple, due to solvation effects, will thus be measured and will therefore be open to interpretation and application.

It is evident that quantitative correlation is possible only between phenomena of exchange of the same ionic species, since the procedure refers to the scale of chemical potential of that species only. For example, we cannot compare quantitatively the acid-base properties, as defined by Brønsted, with solvoacidity properties based on considerations of ionic species other than the hydrogen ion.

As far as experiments are concerned, it appears at

present that correlations of properties have been contemplated essentially in the case of traditional acid-base properties and that efforts concerning exchange phenomena of other ionic species are rare.

We must also give special consideration to the correlation between oxidation-reduction properties.

Correlation between the potential scales and between oxidizing-reducing properties in different solvents

The correspondence between the oxidizing-reducing and electrochemical properties in different solvents is of both theoretical and practical interest, to an extent equal to that between acid-base properties.

1. According to a similar method, it is first important to establish the correspondence between the potential scales in each solvent, with reference to an absolute scale which in this case plays the part of a scale of chemical potential of the electron.

This correlation amounts in practice to determination of an electrode potential in a given medium with regard to a reference electrode, which is invariable and independent of the medium. Firstly, we can consider use of a reference electrode constituted in a definite solvent, for example the normal hydrogen electrode in aqueous solution. But, in order to measure the potential of an electrode in a solvent other than water with regard to this invariable reference standard, there must be a liquid junction between the two media composed of two different solvents; it is known that a junction potential E_j then appears at the junction, which forms part of the overall potential difference measured between the two electrodes. Since the value of the junction potential between two solvents is unknown, we cannot therefore connect the potential of the electrode in the non-aqueous solvent with the potential of the normal hydrogen electrode in water (for which we take $E=0$) more exactly than with an uncertainty corresponding to the value of this junction potential.

In order to minimize the junction potential, and if possible render it calculable, it is necessary to choose a reference electrode comprising a solvent identical to that of the other half-cell. But then, having regard to the variations, due to solvation changes, of the chemical potentials of the dissolved species in the reference oxidizing-reducing system, the reference electrode potential changes with the chosen solvent in a manner which is a priori unknown; this prevents any correlation between different solvents.

To obtain this correlation, it is in principle necessary to have available a reference system with a potential which is invariable whatever the solvent. This problem has already been considered in V.1; we have seen that oxidizing-reducing couples such as Rb^+/rubidium amalgam, or especially ferrocene/ferricinium$^+$, had been suggested as probably satisfying this condition sufficiently nearly. Due to these, it has been possible to establish an approximate correspondence between the accessible potential ranges in different solvents and that in aqueous solutions. Figure 16 shows some examples.

The correlation shows that certain common solvents – such as acetic acid, acetonitrile, propylenecarbonate – permit the attainment of higher potential values than the limit accessible in aqueous solution, and therefore permit the realization of media of greater oxidizing strength. Drastic oxidation processes, impossible in aqueous solution, can thus be made possible by use of such solvents. On the other hand, media of greater reducing strength than those which can be produced from water are also obtainable in certain non-aqueous solvents: DMF, DMSO, HMPA, etc. Drastic reduction processes, impracticable in aqueous solution – such as reduction by alkali metals – become possible in these solutions.

2. After producing this correspondence between the potential scales, we can concern ourselves with determinations of the modifications to the (absolute) oxidizing-reducing strength of a system, which are caused by changes of solvent. Using an argument similar to that used for acid-base couples (relations V-56 to V-59), it is easily shown that the variation of potential on change of solvent (from water to another solvent S) – that is the potential determined with respect to an invariable reference – of one system is connected with the transfer activity coeffi-

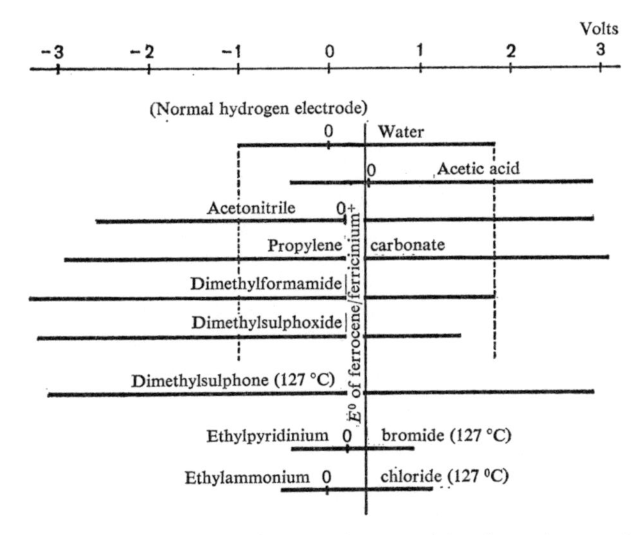

Fig. 16. Correspondence between the potential scales and comparison of the potential ranges available in different molecular solvents, at ordinary temperature and, in three cases (two of these being molten salts), at 127°C. The origin $E=0$ indicated corresponds to the normal hydrogen electrode in the solvent in question.

cients of the oxidizing species Ox and of the reducing species R by:

$$\Delta E = \frac{2.3RT}{nF} [\log \gamma_t(\text{Ox}) - \log \gamma_t(\text{R})]. \tag{V-64}$$

The same couple gives therefore a medium which is more oxidizing in the non-aqueous solvent than in water if $\gamma_t(\text{Ox}) > \gamma_t(\text{R})$, and more reducing in the opposite case.

If $\gamma_t(\mathrm{Ox})=\gamma_t(\mathrm{R})$, that is if solvation of the oxidized form and solvation of the reduced form vary in the same manner, the potential remains invariable and the oxidizing-reducing system can be used as a reference for correlation between potential scales (the hypothesis used in the case of ferrocene, for example).

Amongst the factors which can lead to modifications in the oxidizing-reducing strength, it is still necessary to allow for differences in dielectric constant of the molecular solvents. Having regard to the contribution of the electrostatic effect to the values of the solvation coefficients $\gamma_t(\mathrm{Ox})$ and $\gamma_t(\mathrm{R})$ – at least one of the two forms Ox and R of the couple is ionic, since the loss or gain of an electron leads to a modification of electric charge – we can therefore predict that passing from aqueous solutions to solutions of lower dielectric constant gives rise, between systems of the different types: $\mathrm{Ox/R^-}$, $\mathrm{Ox^+/R}$, $\mathrm{Ox^{2+}/R^+}$, $\mathrm{Ox^{2+}/R}$, $\mathrm{Ox^+/R^-}$, to changes in the standard potentials, more or less appreciable as the species have smaller or larger dimensions. But it must not be forgotten that the more specific solvation effects are very often greater than this electrostatic effect and that the experimentally observed modifications to the oxidizing-reducing strengths therefore correspond much less to the latter than to the former.

BIBLIOGRAPHY

Books on physico-chemical properties in non-aqueous solvents and in molten salts, and on their applications (mostly analytical). In chronological order:

Lorenz, R., *Die Elektrolyse geschmolzener Salze*, Halle, Knapp, 1906.

Walden, P., *Elektrochemie Nichtwässriger Lösungen*, Leipzig, J. A. Barth, 1924.

Jander, G., *Die Chemie in wasserähnlichen Lösungsmitteln*, Berlin, Springer, 1949.

Bell, R. P., *Acids and Bases*, London, Methuen, 1952.

Fritz, J. S., *Acid-Base Titrations in Non-Aqueous Solvents*, G. F. Smith Chem. Co., 1952.

Audrieth, L. F. and Kleinberg J., *Non-Aqueous Solvents*, New York, Wiley, 1953.

Flood, H., Førland, T. and Grjotheim, K., *The Physical Chemistry of Melts*, London, Inst. of Mining and Metall., 1953.

Palit, S. R., Das, M. N. and Somayajulu, G. R., *Non-Aqueous Titrations*, Calcutta, Indian Assoc. for the Cultivation of Science, 1954.

Beckett, A. H. and Tinley, E. H., *Titration in Non-Aqueous Solvents*, BDH Ltd., 1956.

Janz, G. J., *Bibliography on Molten Salts*, Rensselaer Polyt. Inst. Troy 1958; 2nd ed., 1961.

Kreshkov, A. P., *Practical Textbook on Acid-Base Titration in non-Aqueous Media*, Moscow, Chem. Tech. Inst., 1958.

Bell, R. P., *The proton in Chemistry*, Cornell, Univ. Press, 1959.

Bockris, J. O'M, White, J. L. and Mackenzie, J. D., *Physico-Chemical Measurements at high Temperature*, Butterworths, 1959.

Kolthoff, I. M. and Bruckenstein, S., *Acid-Bases in Analytical Chemistry* (*Treatise on Analytical Chemistry*, edited by I. M. Kolthoff and P. J. Elving, Part. I, vol. 1), Interscience, 1959.

West, T. S., Titrations in Non-Aqueous Solvents (in *Comprehensive Analytical Chemistry*, edited by C. L. Wilson and D. W. Wilson, Part I, vol. 1), Interscience, 1959.

Delimarskii, Y. K. and Markov, B. F., *Electrochemistry of Fused Salts*, The Sigma Press, 1961.

Ives, D. J. G. and Janz, G. J. (editors), *Reference Electrodes*, Academic Press, 1961.

Davies, C. W., *Ion Association*, Butterworths, 1962.

Charlot, G. and Trémillon, B., *Les réactions chimiques dans les solvants et les sels fondus*, Gauthier-Villars, 1963; translated into Polish, 1968, translated into English, Pergamon, 1969.

Jander, G., Spandau, H. and Addison, C. C. (editors), *Chemistry in Non-Aqueous Ionizing Solvents*, 5 vol., Vieweg & Pergamon, 1963–1967.

Blander, M., *Molten Salts Chemistry*, Interscience, 1964.

Huber, W., *Titrationen in Nichtwässringen Lösungsmitteln*, Akad. Verlag, 1964; translated into English, Academic Press, 1967.

Lepoutre, G. and Sienko, M. J., *Solutions métal-ammoniac*, Univ. Cath. Lille, 1964.

Sisler, H., *Chemistry in Non-Aqueous Solvents*, Reinhold, 1964.

Sundheim, B. R., *Fused Salts*, McGraw Hill, 1964.

Alabyshev, A. F., Lantratov M. F. and Morachevski, A. G., *Reference Electrodes for Fused Salts*, The Sigma Press, 1965.

Holliday, A. K. and Massey, A. G., *Inorganic Chemistry in Non-Aqueous Solvents*, Pergamon, 1965.

King, E. J., *Acid-base Equilibria*, Pergamon, 1965.

Kucharsky, J. and Safarik L., *Titrations in Non-Aqueous Solvents*, Elsevier, 1965.

Petit, G., *La cryométrie à haute température et ses applications*, Masson, 1965.

Waddington, T. C., *Non-Aqueous Solvent Systems*, Academic Press, 1965.

Lumsden, J., *Thermodynamics of Molten Salt Mixtures*, Academic Press, 1966.

Delahay, P. and Tobias, C. W., *Advances in Electrochemistry and Electrochemical Engineering*, vols. 4 and 5, Interscience, 1966–1967.

Bloom, H., *The Chemistry of Molten Salts*, New York, W. A. Benjamin Inc., 1967.

Gyenes, I., *Titrations in Non-Aqueous Media*, London, Iliffe, 1967. – *Titrationen in nichtwässrigen Medien*, 3rd ed., Stuttgart, F. Enke Verlag, 1970.

Lagowski, J. J. (editor), *The Chemistry of Non-Aqueous Solvents*, 3 vol., Academic Press, 1966–1967, 1970.
Covington, A. K. and Jones, P., *Hydrogen Bonded Solvent Systems*, London, Taylor and Francis, 1968.
Gutmann, V., *Coordination Chemistry in Non-Aqueous Solutions*, Springer Verlag, 1968.
Jander, J. and Lafrenz, C., *Wasserähnliche Lösungsmittel*, Verlag Chemie, 1968.
Zingaro, R. A., *Non-Aqueous Solvents*, D. C. Heath, 1968.
Coetzee, J. F. and Ritchie, C. D., *Solute-Solvent Interactions*, M. Dekker, 1969.
Mamantov, G., *Molten Salts*, M. Dekker, 1969.
Morand, G. and Hladik, J., *Electrochimie des sels fondus*, Masson, 1969.

Collections of physical constants of solvents.

Maryott, A. A. and Smith, E. R., *Table of Dielectric Constants of Pure Liquids*, Washington, NBS Circ. 514, 1951.
Riddick, J. A. and Toops, Jr., E. E., *Organic Solvents. Physical Properties and Methods of Purification* (in *Technique of Organic Chemistry*, edited by A. Weissberger, vol. VII), Interscience, 1955.
Marsden, C., *Solvents Guide*, Cleaven-Hume Press, 1963.
Clark, P. V., *Physical Properties of Fused Salts Mixtures*, 2 vol., Clearing House, 1965-1966.
Janz, G. J., *Molten Salts Handbook*, Academic Press, 1967.

References of papers quoted in the text (*by alphabetical order of authors quoted*)

Alfenaar, M. and de Ligny, C. M., 1965–1967, *Rec. Trav. Chim.* **84**, 81; **86**, 929.
Altar, W., 1937, *J. Chem. Phys.* **5**, 577.
Badoz-Lambling, J., Herlem, M. and Thiebault, A., 1969, *Anal. Letters* **2**, 35.
Barton, J. L., and Bloom, H., 1959, *Trans. Faraday Soc.* **55**, 1792.
Bates, R. G., 1964, *Determination of pH*, New York, J. Wiley, chap. 6
Bates, R. G., 1968, *Hydrogen-Bonded Solvent Systems*, London, (A. K. Covington and P. Jones editors) Taylor and Francis Ltd., p. 49.
Bauer, D., 1965, *Bull. Soc. Chim. France*, 3302.

Bell, R. P., Robinson, R. R., 1961, *Trans. Faraday Soc.* **57**, 965.

Bjerrum, N., 1926, *Kgl. Danske Vidensk. Selsk.* (*Math. fys. Medd.*) **7**, no. 9.

Blander, M., 1961, *J. Chem. Phys.* **34**, 697.

Blander, M., 1964, *Molten Salt Chemistry*, New York, Interscience, chap. 3.

Blander, M., Braunstein, J., 1959–1961, *J. Phys. Chem.* **63**, 1262; *Ann. N.Y. Acad. Sci.* **79**, 838; *J. Chem. Phys.* **34**, 432.

Blander, M., Braunstein J. *et al.*, 1961-1962, *J. Phys. Chem.* **65**, 1866; **66**, 2069; *J. Amer. Chem. Soc.* **84**, 1538, 2028.

Bloom, H. and Bockris, J. O'M, 1964, *Fused Salts*, New York, B. R. Sundheim (editor), McGraw Hill, chap. 1.

Bombara, G., Baudo, G., Tamba, A., 1968, *Corrosion Science*, **8**, 393.

Born, M., 1920, *Z. Physik*, **1**, 45.

Braude, E. A. and Stern E. S., 1948, *J. Chem. Soc.* 1976.

Breant, M. and Demange-Guerin, G., 1969, *Bull. Soc. Chim. France*, 2935.

Bredig, M. A., 1964. *Molten Salt Chemistry*, New York, M. Blander (editor), Interscience, chap. 5.

Bressler, S. E., 1939, *Acta Physicochim. URSS*, **10**, 491.

Bronsted, J. N., 1923, *Rec. Trav. Chim.* **42**, 718. –, 1928, *Chem. Rev.* **5**, 231.

Bronstein, H. R. and Bredig, M. A., 1958, *J. Amer. Chem. Soc.* **80**, 2077.

Bruckenstein, S. and Kolthoff, I. M., 1956, *J. Amer. Chem. Soc.* **78**, 10, 2974.

Bunsen, R. and Kirchhoff, G., 1861, *Pogg. Ann.* **113**, 364.

Cady, H. P. and Elsey, H. M., 1928, *J. Chem. Educ.* **5**, 1425.

Charlot, G. and Trémillon, B., 1963, *Les réactions chimiques dans les solvants et les sels fondus*, Gauthier-Villars, p. 75.

Conte, A. and Ingram, M. D., 1968, *Electrochim. Acta*, **13**, 1551.

Corbett, J. D., 1964, *Fused Salts*, New York, B. R. Sundheim (editor), McGraw Hill, chap. 6.

Coulter, L. V. *et al.*, 1959, *J. Amer. Chem. Soc.* **81**, 2986.

Courtot-Coupez, J. and Le Demezet, M., 1969, *Bull. Soc. Chim. Frame*, 1033.

Cubicciotti, D., 1952, *J. Amer. Chem. Soc.* **74**, 1198.

Davies, C. W., 1962, *Ion Association*, London, Butterworths, p. 168.

Davis, M. M., 1968, *Acid-Base Behavior in Aprotic Organic Solvents*, NBS Monograph 105.
Dawson, L. R., and Wharton, W. W., 1960, *J. Electrochem. Soc.* **107**, 710.
Debye, P. and Hückel, E., 1923, *Physik Z.* **24**, 185, 305.
Delarue, G. 1960, *Bull. Soc. Chim. France*, 906, 1654.
Denison, J. T. and Ramsey, J. B., 1955, *J. Amer. Chem. Soc.* **77**,2615.
Devynck, J. and Trémillon, B., 1969-1971, *J. Electroanal. Chem.* **23**, 241; **30**, 443.
Devynck, J., 1971, Thèse de doctorat ès sciences, Paris.
Douheret, G., 1967, *Bull. Soc. Chim. France*, 1412, 2969.
Drago, R. S. and Meek, D. W., 1961, *J. Phys. Chem.* **65**, 1446; *J. Amer. Chem. Soc.* **83**, 4322.
Duke, F. R., and Iverson, M. L., 1958–1959, *J. Phys. Chem.* **62**, 417; *J. Amer. Chem. Soc.* **80**, 5061; *Anal. Chem.* **31**, 1233.
Durand, G., 1970, *Bull. Soc. Chim. France*, 1220.
Evert, L. and Konopik, N., 1949, *Öst. Chem. Ztg.* **50**, 184.
Eluard, A., 1970, Thèse de doctorat ès sciences, Paris. Eluard, A. and Trémillon B., 1968–1970, *J. Electronal. Chem.* **18**, 277; **26**, 259; **27**, 117.
Eyring, H. *et al.*, 1937, *J. Phys. Chem.* **41**, 249. –, 1958, *Proc. Nat. Acad. Sci.* **44**, 683.
Feakins, D. *et al.*, 1963, *J. Chem. Soc.* 4734. –, 1966, *J. Chem. Soc.* 714.
Flood, H. and Førland, T., 1947, *Acta Chem. Scand.* **1**, 592.
Flood, H., Førland, T. and Roald, B. 1947, *Acta Chem. Scand.* **1**, 790.
Flood, H., Førland T. and Grjotheim, K., 1953, *The Physical Chemistry of Melts*, London, Methuen. – 1954, *Z. anorg. allg. Chem.* **276**, 289.
Førland, T., 1955, *J. Phys. Chem.* **59**, 152.
Førland, T., 1957, *Norg. Tek. Vitenskapsakad*, Ser. 2, no. 4.
Frank, H. S. and Wen, W. Y., 1957, *Disc. Faraday Soc.* **24**, 133.
Frank, W. B. and Foster, L. M., 1957, *J. Phys. Chem.* **61**, 1531.
Franklin, E. C., 1953, *The Nitrogen System of Compounds*, New York, Reinhold.
Franks, F. and Ives, D. J. G., 1966, *Quart. Rev.* **20**, 1.
Frenkel, Y. I., 1935, *Acta Physicochim. URSS*, **3**, 633, 913.
Fuoss, R. M., 1958, *J. Amer. Chem. Soc.* **80**, 5059.

Fürth, R., 1941, *Proc. Cambridge Phil. Soc.* **37**, 252, 276, 281.
Germann, A. F. O., 1925, *J. Amer. Chem. Soc.* **47**, 2461.
Gilkerson, W. R., 1956, *J. Chem. Phys.* **25**, 1199.
Gillespie, R. J., 1950, *J. Chem. Soc.* 2473, 2493, 2516.
Goret, J. and Trémillon, B., 1966, *Bull. Soc. Chim. France*, 67, 2872.– 1967, *Electrochim. Acta*, **12**, 1065.
Grall, M., 1966, Rapport CEA-R 1902.
Grunwald, E., 1962, *Electrolytes*, B. Pesce (editor), Oxford, Pergamon, p. 62.
Grunwald, E., and Berkowitz, B. J., 1951, *J. Amer. Chem. Soc.* **73**, 4939.
Grunwald, E., *et al.*, 1958–1960, *J. Amer. Chem. Soc.* **80**, 3840; **82**, 5801.
Guggenheim, E. A., 1929, *J. Phys. Chem.* **33**, 842.
Guiot, S. and Trémillon, B., 1968, *J. Electroanal. Chem.* **18**, 261.
Gutbezahl, B. and Grunwald, E., 1953, *J. Amer. Chem. Soc.* **75**, 559.
Gutmann, V., 1952, *Monatsh.* **83**, 164.
Gutmann, V., 1968, *Coordination Chemistry in Non-Aqueous Solutions*, Vienna, Springer Verlag.
Gutmann, V., Baaz, M., *et al.*, 1959–1961, *J. Phys. Chem.* **63**, 378; *Monatsh.* **90**, 239, 256, 271, 276, 426, 729; **92**, 272, 707, 714; *J. Inorg. Nucl. Chem.* **18**, 276.
Gutmann, V. and Lindqvist, I., 1954, *Z. physik. Chem.* **203**, 250.
Gutmann, V. and Wychera, E., 1965, *Inorg. Nucl. Chem. Letters*, **2**, 257.
Halle, J. C., Gaboriaud, R., and Schaal, R., 1969, *Bull. Soc. Chim. France*, 1851.
Hammett, L. P., 1935, *Chem. Revs.* **16**, 67.
Hammett, L. P. and Deyrup, A. J., 1932, *J. Amer. Chem. Soc.* **54**, 2721, 4239.
Harned, H. S., 1925, *J. Amer. Chem. Soc.* **47**, 930.
Harned, H. S. and Fallon, L. D., 1939, *J. Amer. Chem. Soc.* **61**, 2371.
Harned, H. S. and Hamer, W. J., 1933, *J. Amer. Chem. Soc.* **55**, 2194.
Harned, H. S. and Owen, B. B., 1950, *The Physical Chemistry of Electrolytic Solutions*, Reinhold, 2nd ed.
Hartley, H. and Buckley, P. S., 1929, *Phil. Mag.* **8**, 320.
Hartley, H. and McFarlane, A., 1932, *Phil. Mag.* **13**, 425.
Herlem, M., 1967, *Bull. Soc. Chim. France*, 1687.

Ingram, M. D., and Janz, G. J., 1965, *Electrochim. Acta*, **13**, 495.
Izmailov, N. A., 1959–1960, *Dokl. Akad. Nauk SSSR*, **126**, 1033; *Zhur. Fiz. Khim.* **34**, 2414.
Izmailov, N. A., 1963, *Dokl. Akad. Nauk SSSR*, **149**, 1103, 1364.
Jander, G., Rüsberg, E., and Schmidt, H., 1948, *Z. anorg. allg. Chem.* 238.
Jander, G. and Schmidt, H., 1943, *Wien Chem. Ztg.* **46**, 49.
Kilpatrick, M. *et al.*, 1940, *J. Chem. Soc.* **62**, 3047, 3051. –, 1953, *J. Electrochem. Soc.* **100**, 185. –, 1956, *J. Chem. Amer. Soc.* **78**, 5183.
Kirkwood, J. G., 1935–1936, *J. Chem. Phys.* **3**, 300; *Chem. Revs.* **19**, 275.
Kolthoff, I. M. and Bruckenstein, S., 1959, *Treatise on Analytical Chemistry*, I. M. Kolthoff and P. J. Elving (editors), Interscience, Part. I, vol. 1, p. 499.
Kolthoff, I. M., Bruckenstein, S., and Chantooni, M. K., 1961, *J. Amer. Chem. Soc.* **83**, 3927.
Kolthoff, I. M., Lingane, J. J. and Larson, W. D., 1938, *J. Amer. Chem. Soc.* **60**, 2512.
Kolthoff, I. M., Stocesoca, D., and Lee, T. S., 1953, *J. Amer. Chem. Soc.* **75**, 1834.
Kraus, C. A., 1921, *J. Amer. Chem. Soc.* **43**, 749. –, 1922, *Properties of Electrically Conducting Systems*, AC.S. Monograph no. 7.
Le Ber, F., 1970, Thèse de doctorat ès sciences, Paris.
Le Port, L., 1966, Rapport CEA-R no. 2904.
Leroy, M., 1962, *Bull. Soc. Chim. France*, 968.
Lewis, G. N., 1938, *J. Franklin Inst.* **226**, 293.
Littlewood, R. and Edeleanu, C., 1960–1962, *Electrochim. Acta*, **3**, 195; *Silicates Indust.* **26**, 447; *J. Electrochem. Soc.* **109**, 525.
Lowry, T. M., 1923–1924, *Chem. and Ind. (London)*, **42**, 43; *Trans. Faraday Soc.* **20**, 13.
Luder, W. F. and Zuffanti, S., 1946, *The Electronic Theory of Acids and Bases*, Wiley.
Lumsden, J., 1961, *Disc. Faraday Soc.* **32**, 138.
Lux, H., 1939, *Z. Elektrochem.* **45**, 303.
Lux, H., 1948–1949, *Z. Elektrochem.* **52**, 220, 224; **53**, 41, 43, 45.
Madic, C. and Trémillon, B., 1968, *Bull. Soc. Chim. France*, 1634.
Marple, L. W., 1969, *J. Chem. Soc.* **17** (A), 2626.

Molina, R., 1961, *Bull. Soc. Chim. France*, 301, 1001, 1184.

Mott, N. F. and Gurney, R. W., 1957, *Electronic Processes in Ionic Crystals*, New York, Oxford Univ. Press, p. 63.

Mukherjee, L. M., 1957, *J. Amer. Chem. Soc.* **79**, 4040.

Nelson, I. V. and Iwamoto, R. T., 1961, *Anal. Chem.* **33**, 1795.

Onsager, L., 1926–1927, *Physik Z.* **27**, 388; **28**, 277.

Parker, A. J., 1962, *Quart. Rev.* (*London*), **16**, 163.

Parker, A. J., *et al.*, 1966–1967, *J. Chem. Soc.* 220; *J. Amer. Chem. Soc.* **89**, 3703, 5549.

Paul, M. A. and Long, F. A., 1957, *Chem. Revs.* **57**, 1.

Perkovets, V. D. and Kryukov, P. A., 1969, *Izv. Sib. Otb. Akad. Nauk SSSR*, Ser. Khim. Nauk, 9.

Picard, G. and Vedel, J., 1969, *Bull. Soc. Chim. France*, 2557.

Pleskov, V. A., 1947, *Usp. Khim.* **16**, 254.

Popovych, O., 1966, *Anal. Chem.* **38**, 558.

Popovych, O. and Dill, A. J., 1969, *Anal. Chem.* **41**, 456.

Pourbaix, M., 1949, *Thermodynamics of Dilute Aqueous Solutions*, London, Arnold.

Pourbaix, M., 1963, *Atlas d'équilibres électrochimiques à 25 °C*, Paris, Gauthier-Villars.

Price, E., 1966, *The Chemistry of Non-Aqueous Solvents*, J. J. Lagowski (editor), New York, Academic Press, chap. 2.

Pucci, A. E., 1968. Thèse de doctorat d'université, Paris.

Pucci, A. E., Vedel, J., and Trémillon, B., 1969, *J. Electroanal. Chem.* **22**, 253.

Rahmel, A., 1968, *Electrochim. Acta*, **13**, 495.,

Reynaud, R., 1967, *Bull. Soc. Chim. France*, 4597.

Romberg, E. and Cruse, K., 1959, *Z. Elektrochem.* **63**, 404.

Rothstein, J., 1955, *J. Chem. Phys.* **23**, 218.

Rumeau, M., 1971, Thèse de doctorat ès sciences, Paris.

Samoilov, O. Y., 1965, *Structure of Aqueous Electrolyte Solutions*, New York, Consultants Bureau.

Schaal, R., Gadet, C., and Souchay, P., 1960, *C. R. Acad. Sci.* **251**, 2529.

Schaal, R,, and Masure, F., 1956. *Bull. Soc. Chim. France*, 1141.

Schaal, R., and Teze, A., 1961, *Bull. Soc. Chim. France*, 1783.

Schaap, W. B., Brewster, P. W., and Schmidt, F. C., 1961, *J. Phys. Chem.* **65**, 990.

Sørensen, S. P. L., 1909, *Biochem. Z.* **21**, 131, 201; *C. R. trav. lab. Carlsberg*, **8**, 1.
Sørensen, S. P. L. and Linderstrøm-Lang, K., 1924, *C. R. trav. lab. Carlsberg* **15**, no. 6.
Stewart, G. W., 1927, *Phys. Rev.* **30**, 232.
Strehlow, H., 1952, *Z. Elektrochem.* **56**, 827. –, 1966, *The Chemistry of Non-Aqueous Solvents*, New York, J. J. Lagowski, (editor), Academic Press, vol. 1, chap. 4.
Strehlow, H. *et al.*, 1960–1961, *Z. Elektrochem.* **64**, 483; *Z. Physik. Chem. N.F.* **30**, 141.
Temkin, M., 1945, *Acta physicochim. URSS*, **20**, 411.
Texier, P., 1968. *Bull. Soc. Chim. France*, 4716.
Touller, J. C., and Trémillon, B., 1970, *Anal. Chim. Acta*, **49**, 115.
Trémillon, B., 1960, *Bull. Soc. Chim. France*, 1940; 1963, *Les réactions chimiques dans les solvants et les sels fondus* (G. Charlot and B. Trémillon, joint authors), Paris, Gauthier-Villars, chap. I and p. 211.
Trémillon, B., Barraqué, C., and Vedel, J., 1968, *Bull. Soc. Chim. France*, 3421.
Trémillon, B. and Létisse, G., 1968, *J. Electroanal. Chem.* **17**, 371, 387.
Trémillon, B. *et al.*, 1965, *Bull. Soc. Chim. France*, 1853.
Tur'Yan, Y. A. *et al.*, 1959–1960, *Zhur. Neorg. Khim.* **4**, 599, 808, 813, 1070; **5**, 1748.
Ussanovich, M., 1939, *Zhur. Obsh. Khim. USSR*, **9**, 182.
Vedel, J., 1967, *Ann. Chimie*, **2**, 335.
Vedel, J., 1968, *Bull. Soc. Chim. France*, 3069.
Vedel, J. and Postel, M., 1970, *J. Electroanal. Chem.*
Vedel, J. and Trémillon, B., 1966, *Bull. Soc. Chim. France*, 220.
Verhoek, F. M., 1936, *J. Amer. Chem. Soc.* **58**, 2577.
Vieland, L. J. and Seward, R. P., 1955, *J. Phys. Chem.* **59**, 466.
Weyl, W., 1863–1864, *Pogg. Ann.* **121**, 601, 697; *Ann. Physik.* **121**, 601.